MAR 1 6

How TO·WIN GAMES ·AND·BEAT· ·PEOPLE·

TOM WHIPPLE

DEY ST.

AN IMPRINT OF
WILLIAM MORROW *PUBLISHERS*

DEY ST.

First published in Great Britain by Virgin Books in 2015

HarperCollins books may be purchased for educational, business,
or sales promotional use. For information please e-mail the Special
Markets Department at SpecialMarkets@harpercollins.com.

HarperCollins Publishers, 195 Broadway, New York, NY 10007.

FIRST DEY STREET BOOKS EDITION 2015

Library of Congress Cataloging-in-Publication Data has been applied for.

ISBN 978-0-06-244374-8

15 16 17 18 19 DIX/RRD 10 9 8 7 6 5 4 3 2 1

For **Mum**, for all the things mums do that no one notices.
For **Catherine**, for showing me what those things are.
For **Dad**, for always letting me win.

CONTENTS

INTRODUCTION

As everyone knows, it's not the winning that counts: it's the taking part. Nonsense! That is the battle cry of the loser. This is not a book for people who enjoy taking part; its readers are not people who, say, bring out a board game on Thanksgiving Day anticipating some pleasant after-dinner fun with the extended family over a bottle of sherry. Rather, this is a book for people who bring out a board game on Thanksgiving Day anticipating a protracted struggle that will act as a proxy conflict for long-simmering family feuds, and which will ultimately end only with the creation of more feuds or, at best, the upending of the board by a belligerent uncle.

This is a book that tells you how to win, put together after consulting the people who know. What is the best way to play Operation? Why not consult a surgeon? How do you construct the ultimate Jenga tower? Ask a structural engineer. Is there a strategy for guaranteeing victory at Connect 4? Let's follow the lead of noted player Beyoncé, and consult with the man who wrote an entire thesis on it.

And for those who say this is against the spirit of the thing, well, Risk does not describe itself as the "the game of taking part and having fun, because really that's what matters." It describes itself as "the game of world domination," because domination is how you win. (And, since you ask, in Risk that requires understanding the probabilities derived from a branch of statistics known as Markov chains, in a paper produced by a mathematician with a little too much time on his hands.)

What, though, if in your desire to win, in explaining to your opponents that you have memorized a 10 x 10 table of probabilities just so you can recapture Yakutsk, it all goes a bit like Napoleon's retreat from Moscow and becomes a messy brawl, possibly involving pillows? Well, this book can help there, too. As Special Forces hero Andy McNab explains, victory "in a pillow-fight situation" is about shock and awe: "The pillow is just a weapon. If you've got the weapon and you're there, you just get in straightaway."

So, if you are the sort of person who buys new trainers for the dads' race at school sports day, do not be ashamed. If you consult Sun Tzu before playing Risk, take heart. If you know that *qi* is a legitimate Scrabble word but you do not know what it means, do not feel alone. Instead, read on. Bad losers, we salute you.

MONOPOLY

WHAT'S IT ALL ABOUT?

A timeless classic: everyone's favorite brutal capitalist board game. Build a powerful property empire and family rifts that will persist for decades. Learn the importance of fiscal prudence, and of squeezing your opponents for everything they are worth. Oh, and win $10 in a beauty contest.

HOW DO YOU PLAY?

Much like in real capitalism, success is a mixture of luck, judgment, and the extent of one's willingness to mercilessly crush the weak. Should you move around the board and buy properties selectively, to develop an exclusive portfolio of luxury residences, or purchase everything you land on and become an overleveraged subprime slum landlord? The choice is yours. Just remember, it might be Christmas but that's still no reason to go easy on Aunt Joan.

HOW DOES IT END?

With the losing player turning over the board or—a true capitalist simulation—going on strike.

ANALYSIS

Buy orange. If John Haigh has one tip for Monopoly players, it is that. A game theorist, John is unusual in that he has actually turned his mathematical attentions toward real games rather than, say, the financial markets. And he says that the most salient fact about Monopoly is not the cost of Boardwalk, the multiplicative value of owning all the stations, or the putative luck value of having the boot, it is that people go to prison.

"The single square that is landed on most often is jail," says John. This is because there are so many ways of arriving at it—to land on Boardwalk you have to throw the correct numbers with the dice, but for jail, "You can hit it naturally; or hit the Go to Jail; or throw three doubles in succession."

Ordinarily, people who have been incarcerated are not safe financial bets. No one, for instance, is making investments on the basis of Kenneth Lay's movements. Not so in Monopoly. It is not so much that one is sent to jail—it is that one leaves it. What comes next is what is important in terms of buying: where do people setting out from jail land? Well, the most common numbers thrown with two dice are 5, 6, 7, 8, and 9. So it seems the orange properties—6, 8, and 9 throws away—offer the steadiest revenue stream.

John's computer model (yes, he has indeed built one) confirms this: "For every 100 hits on purple or blue, you tend to get 110 on green or yellow, and 122 on orange or red."

As every decent profiteering landlord will tell you, though, it is not just about the frequency with which you can take rents, it is also about the amount you can extract. On this measure, too, the oranges do well. You may get more from renting a bijou flat in Manhattan to a Russian oligarch, but it costs you more, too. In Monopoly, as in life, there is a happy medium between an efficiency on Mediterranean Avenue and a penthouse on Park Place. "Add up the total required to buy all the properties and put hotels on them," says John. "Then add up the maximum rent on each property. The higher the ratio of income to cost, the more attractive the set is to own. On this measure, light blue is best at 1.59, then orange (1.41), deep blue (1.27), purple (1.24), yellow (1.15), brown (1.13), red (1.09) and finally green (1.01)."

Assuming you've beaten your opponents to the best property, and achieved the optimal cost-to-income ratio, what next? Well, it is time to crush them. Here, too, it is not the high-rolling glamour of Boardwalk that is your friend, but the steady, dependable mediocrity of St. James Place.

How, asks John, do you most quickly get to the situation where you can extract rents "large enough to cause embarrassment"? Let us assume that $750 is a sufficient sum to bankrupt Uncle Simon after a few glasses of sherry, what then is the minimum for each set of properties that we can spend to achieve that?

With brown and light blue you can't get this sum; for the rest you can get there by spending $1,760 (orange), $1,940 (purple), $1,950 (deep blue), $2,150 (yellow), $2,330 (red) or $2,720 (green).

Of course, all this highlights is that this isn't really a perfect simulation of the capitalist market at all. Orange is clearly undervalued without a way of correcting the price—an economic problem that is compounded by the game's shameless quantitative easing in the form of $200 every time you pass Go. Real Monopoly players have long known this, which is why a serious game will see side deals in the form of offers to do the washing-up, risk-offsetting in the form of promises to take the dog for a walk and, ultimately, capital flight into stabler currencies. Such as Russell Stover chocolates.

Isn't the free market wonderful?

"Monopoly," says Mark Littlewood, waving a cigarette in one hand while bashing a pint on the table for emphasis with the other, "is a vicious North Korean fight to get your hands on other people's property."

Doesn't it represent the ugly side of unfettered capitalism? "Absolutely not. It is intrinsically anti-free market."

Mark is director general of Britain's Institute of Economic Affairs. A strident opponent of the state, of socialism, of the minimum wage for disabled people (and, in fairness, for everyone else), he doesn't buy the idea that Monopoly epitomizes the grand capitalist struggle. Quite the reverse. Instead, it is a communist fifth column.

"The problem is, it's a zero-sum game: if I win, you lose. A free market is not a zero-sum game. In the real world almost everyone leaves the Monopoly table a bit richer. There is no growth mechanism in the game. No one ever says, as they do in life, 'Oh, I've got Mediterranean Avenue now. I'll retire, watch TV, and live off my pension.'"

He goes on, invoking Adam Smith. "The size of the pie in Monopoly is fixed, but what free markets do is expand the pie. Last time I checked, Boardwalk was worth more than $400." At one point, though, it was, and the fact it is not now, says Mark (and Adam Smith) "isn't just because

[you] get $200 every time they pass Go," it is because, ultimately, capitalism makes everyone richer.

So how do you create a Monopoly that is capitalist? Well, as a start you open it up to the idea that, as in the market, anything can be bought—including your position on the board. "If you had the situation whereby you said the winner would be given $500 that is sitting under the Christmas tree, things would be very different.

"You would then get people trading in their position in Monopoly against the $500 under the tree." For those unlikely to find $500 under the Christmas tree, a box of chocolates works just as well—or, indeed, anything else that is divisible and valuable.

In Mark's version of the game people could make an assessment of their position as they go along and decide, say, that they have a 50 percent chance of winning so it is worth cashing out and selling their properties to an opponent for $250. "Or even," Mark suggests, "they might say, 'I'm well ahead, but I don't want to die of boredom over the next four hours. I'll take the $250 and divide the rest between everyone else, just so that the bloody thing can end.'"

If Monopoly sometimes feels unfair, if success or failure seems arbitrary, that may be because it is meant to be. Monopoly was created in the early twentieth century by a woman named Elizabeth Magie, and in its original guise as the "Landlord's Game" it was designed to demonstrate the benefits of "Georgism." This was an economic analysis named after Henry George, a man who believed that unfair benefits accrue to landowners as they can extract money from owning their property without ever adding value or improving society through it.

Writing in the 1902 edition of the (much-missed) *Single Tax Review*, Ms. Magie wrote of her game, "It is a practical demonstration of the present system of land-grabbing with all its usual outcomes and consequences. It might well have been called the Game of Life, as it contains all the elements of success and failure in the real world, and the object is the same as the human race in general seems to have, i.e., the accumulation of wealth."

FACT

During World War II, Allied soldiers were sent Monopoly games in the mail—but with one small change: the money inside was real German and Italian currency, the playing pieces concealed compasses and files, and the board hid a printed silk map.

PILLOW
FIGHTING

WHAT'S IT ALL ABOUT?

As Clausewitz should really have said, "Pillow fighting is the continuation of war by other means." As Sun Tzu did say, "Let your plans be dark and impenetrable as night, and when you move, fall like a thunderbolt." And never use one of those cheap foam pillows, for they have no weight behind them.

HOW DO YOU PLAY?

First, find a pillow. Next, hit someone with it. Repeat.

HOW DOES IT END?

With a lot of feathers, a bit less usable bedding, and, if you use one of those orthopedic pillows, an overwhelming victory only slightly tempered by the mild concussions of your opponents.

ANALYSIS

"All you are," says Andy McNab, in the kind of voice Jack Bauer uses while searching for the clamp with which to attach electrodes to a terrorist's nipples, "is a platform for the weapon. It doesn't matter whether it is a rocket launcher or a pistol, if the platform isn't right the weapon won't do what it is supposed to do."

Andy—Special Forces hero, private security contractor, veteran of covert operations around the world, victim of torture after being captured in the first Gulf War—has been in a lot of fights, and won a lot of fights. Pillow fights, to his mind, differ only in the fact that they involve pillows. Rather than, say, choke holds, shrapnel, and high-velocity 7.62 bullets. Oh, and also in that, ideally, he explains, "You shouldn't really bite, gouge their eyes, or grab their bollocks."

"The pillow is just a weapon," he says. "The principles are the same as for any weapon: your position and hold must be firm enough to support it. The chances are you are going to have it in one hand, so you can have the other hand free to push the target away from you, or to grab the target toward you and control it." He advocates a stable stance with feet set firm and apart. "The next principle is that the weapon must point naturally at the target. It's no good being in a great stable position if the target is behind you. When you swing the pillow you need maximum damage with minimum effort. You won't hit the target with maximum force if you have to use some of that force to move your body around."

Andy believes his chief advantage, though, in a really serious pillow fight situation, would be the willingness to go in fast and hard with overwhelming force. He is a firm advocate of the shock and awe approach to late-night dormitory squabbles.

"The fact is, once you commit, you commit. Culturally, from films, we are conditioned to think you've got to dance around first. But once you're in a situation, the principle is there's no turning back. It's that recognition that you've got to fight, or you're in the shit—so you get on and fight. If you've got the weapon and you're there, you just get in straightaway." Or, just go for their bollocks.

FACT

In 1897 *The Times* of London first mentions pillow fighting in its report of a divorce proceeding. The husband conceded, when asked about the couple's turbulent relationship, that they "used frequently to have pillow fights."

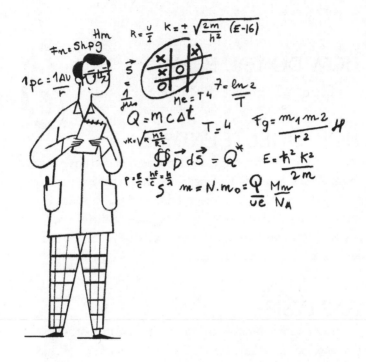

TIC-TAC-TOE

WHAT'S IT ALL ABOUT?

The Romans played it. So, too, almost certainly, did the Ancient Egyptians. Which is hardly surprising since the prerequisites are just sand and boredom.

HOW DO YOU PLAY?

By finding a very, very stupid opponent.

HOW DOES IT END?

In an unsatisfying impasse. If it doesn't end in an unsatisfying impasse then one of two things has happened. 1. You lost—in which case it seems doubtful you are competent enough to read this. 2. Your opponent lost—in which case hopefully they are not someone you consider a good friend and/or your intellectual equal.

ANALYSIS

Most reasonable assessments would have to conclude that one notorious American competitor, named Ginger, has an unfair, to the point of being insurmountable, advantage in tic-tac-toe. Traveling the country's casinos, gamblers line up to play her, but while they only win if they beat her, she gets to keep their money even if there is a draw.

Given that two half-decent tic-tac-toe players always draw, the results are hardly surprising. "Usually we will play about 500 people a day," says Kelly Boger, Ginger's manager. "We would maybe expect one or two losses."

There is only one detail that might explain why gamblers continue to believe they might have an edge. Ginger is a chicken.

In fact, Ginger is the latest in a long line of chickens that has been enriching the Boger family of animal trainers for the past twenty years.

"My dad started training animals years and years ago for circuses," says Kelly. "Buffaloes, camels, mountain lions, giraffes." As a sideline, Kelly trained up some chickens—getting them to peck at the right squares in tic-tac-toe—and began taking them around to fairs. He didn't think much of it initially. "It was just a quarter a play. Then I got some chickens stolen when I was in Pennsylvania." That turned out to be Kelly's big break. "It went on national news, and I started getting these calls from casinos. So I had to train some more."

Since then, Kelly has trained several hundred chickens—and has the process down to a fine art. "It's positive reinforcement. If they do something correct, you give them a reward for it. Then it just takes a lot of patience, and a lot of time." Is there anything that human players can learn from his training regime? "They pick a middle or a corner if they go first," he says of his trained chickens. "That's where you've got to start. Then you get them to start strategizing where the person's going to go versus where they are actually going to go."

Basic tic-tac-toe strategy is simple. The diagonals—the corners and the center—are most valuable. Three different lines can start or end on a corner square, whereas the four squares not on a diagonal can only be part of two lines.

Three lines possible from one corner

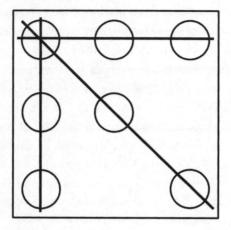

Just two lines possible from one edge

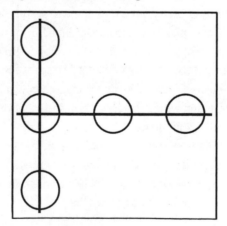

So, initially, the best move is a center or a corner. If the center isn't taken by the first player, it should be by the second. After that, each move is either about blocking any threats from your opponent or, if there aren't any, making threats of your own. Ideally, you will look to get a "fork," where on the next go you are threatening two separate lines.

Fork

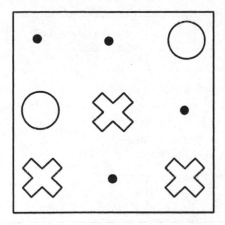

It may not be the most complex of games, but isn't it still a little humiliating for humans that something as stupid as a chicken can master it? Kelly has a correction: "Maybe when I first get 'em they're stupid, but not when I'm done with them. It's like you," says Kelly. "I could get you to do a lot of things for food."

START[+]

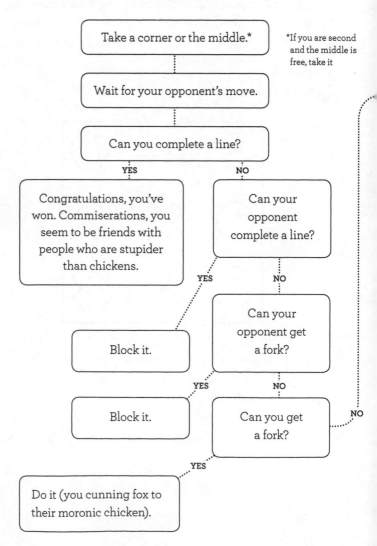

Take a corner or the middle.*

*If you are second and the middle is free, take it

Wait for your opponent's move.

Can you complete a line?

YES

Congratulations, you've won. Commiserations, you seem to be friends with people who are stupider than chickens.

NO

Can your opponent complete a line?

YES

Block it.

NO

Can your opponent get a fork?

YES

Block it.

NO

Can you get a fork?

NO

YES

Do it (you cunning fox to their moronic chicken).

Back we go to the beginning. But really we all know this is heading for an unsatisfactory draw. Why not try something involving more skill, like scissors, paper, stone, or roshambo?

Are there any diagonals left? ········ **NO** ········ I suppose you can keep playing

FACT

Tic-tac-toe was probably the first game to be "solved" by a computer. It is trillions of times simpler than some of the games solved today, but even so there are 26,830 potential positions.

CHARADES

WHAT'S IT ALL ABOUT?

Embarrassing yourself, through the medium of mime.

HOW DO YOU PLAY?

Choose a common cultural reference, then act it out in a way that makes it obvious to everyone what it is. Or doesn't. Best to avoid, for instance, *À la recherche du temps perdu*.

HOW DOES IT END?

When either the booze runs out, the guests pass out, or you stop explaining to anyone who will listen that "It was clearly a bloody madeleine. Go on then, you show me how you'd do a madeleine. I thought not. That's a terrible madeleine."

ANALYSIS

When Nola Rae plays charades with her friends, she does not just restrict herself to the conventional book/film/play formalism. "I've seen someone do a mime about a tea bag being murdered," she says. "Another of my friends did a bowling ball. She picked herself up by her nose." Then again, as cofounder of the London International Mime Festival, Nola does have some friends who are rather good at charades.

That should be no excuse, though. In Nola's opinion, we all have a mime artist in us—and with charades, all we have to fear is fear itself. And, maybe, sobriety. "You just have to bring out the best mime in you. Be as simple as possible, and don't flap—too many movements are confusing. Charades is about imagination—yours, and the audience's."

It is also about finding better ways to direct that flapping. A vigorous waving motion accompanied by a mute squawk— a little like a fat turkey trying to fly—might be the universally accepted mime for "You're really close, but not quite there … you total jerk," but experienced players know how to communicate even better.

Charades guides are a thing to behold; among aficionados there exists what essentially constitutes a full sign language, complete with Chomskian grammar. The largest such document online ends with the rather optimistic "relax and have fun." But there are things that can be gleaned from these guides. If the guess is correct, but should be in the past tense, move a hand backward over your shoulder. If it is the opposite of what was guessed, hold your palms in front of you and swap your hands over. If someone needs to be more specific—for instance, guessing "philosopher" when they need to guess "Wittgenstein"—grind a fist into one palm. Language, after all, should not be the limits of one's understanding.

What, though, if no one else knows those symbols? You don't want someone shouting out "pestle and mortar" when your fist-grinding motion actually means, "Well done, 'red wine' is a good guess, but I actually need the *appellation contrôlée* as well." After all, charades may be an opportunity to prove your superior acting skills, but the game remains—at least ostensibly—a collaborative experience. This is something that gets increasingly frustrating if a slightly tipsy Aunt Margaret keeps on asking, "How do I do a book again?" So make sure to drill people in symbols before the port arrives and then, the formalism learned, they can fully appreciate the majesty of your imaginative genius—even if they might not appreciate why they invited you in the first place.

Nola is very keen to unleash that genius, too. "I suggest at Christmas you don't do the usual charades. Try Christmas things—do a turkey, or a tree. If you're stuck, having a bit of booze helps.

"Really, though, you can mime almost anything." She pauses. A slight correction. "I always remember a girl trying to mime a pink banana. It didn't go well—you can't mime colors unless you point, which is cheating."

FACT

History's first recorded mime artist was probably Telestes, a dancer employed by the Greek playwright Aeschylus, who was noted for his ability to act out scenes using his body. Aeschylus died when a tortoise was dropped on his head by an eagle—a scene that even Telestes might have had trouble conveying in charades.

JENGA

WHAT'S IT ALL ABOUT?

Do you like to use the word *cantilever*? Do people buy you wall calendars of bridges at Christmas? Have you ever made a model steam train? Then this is probably the game for you.

HOW DO YOU PLAY?

Demonstrate your affinity with the basic engineering principles behind today's great buildings by constructing a perfect hollow tower, with each move precisely counteracting and stabilizing that of your opponent. Maybe talk about "force vectors." Inevitably, you will instead end up demonstrating your affinity with the basic engineering principles behind the Leaning Tower of Pisa.

HOW DOES IT END?

Messily, with a clatter.

ANALYSIS

Leslie Scott has a very good response to those Jenga fans who call her a cheat. Which is lucky because, with the tactics she uses, she really needs one. "Imagine you've got a layer in which the middle block has been removed," she says. It is a situation with which any experienced player who takes on amateurs will sympathize—only novices would be so inefficient as to leave a row requiring two blocks to remain. Her approach to this is probably not one, however, with which they will identify so strongly. "What I like to do is very, very gently squeeze together the two remaining blocks with my thumb and forefinger." In this way, she can shift one to the middle, and remove the other.

The heresy continues. "There's another thing I do, which people say they are sure is not allowed. To steady the tower when I am removing a block I put my elbow on the table and rest my forearm vertically against it." In this way she can pull harder on the blocks, confident the tower can't topple.

So long as the hand removing the block is also the one attached to that forearm, this is, she claims, perfectly legal. "It's legitimate—people have this idea you must have your elbow out in midair while you delicately remove the blocks. That's not true."

And what if her opponents object to this rather loose interpretation of the rules? Well, she just tells them that she must be correct—after all, it was she who wrote those rules.

Jenga began life as something Leslie played with her younger brother, when she was eighteen and he was five. "He had these handmade wooden playing blocks, and we used to pile them on top of each other." Eventually the game developed from random piling to something more methodical. It also became something that her friends wanted to play as well. At some point "the penny dropped" that she could potentially make a business out of it.

Commercializing it proved difficult, though—not least because the secret to making, and playing, Jenga is that it relies on imperfections of just the sort that mass production eliminates. If every block were precisely the same, then Jenga would be a very different game. Instead of scanning the tower for loose blocks, players would find each block fitted snugly and just as hard to remove as the others.

To get around this, before the blocks are cut from a plank of wood, that wood is passed through a sander whose surface changes in height ever so slightly. The blocks are then "tumble polished" in a barrel—a technique that adds another layer of randomness. Finally, they are boxed up in groups of 54—each set entirely unique—and sent out with one of the simplest sets of instructions of any commercial game.

"There are only three rules to this game," says Leslie. "You must use only one hand; you can't take out from the penultimate row until the top row is complete; and if a brick hits the floor, you've lost." It was this latter rule that caught out one of her correspondents.

"I had the sweetest email from someone the other day. He said he used a particular technique all the time and wanted to know if it was legal. He sent me a video." The video showed him flicking out the blocks at speed, like a conjurer pulling a tablecloth from under crockery. "I had to say 'I'm sorry, it's not allowed'—it involved the brick having to hit the ground."

Although she did also suggest to him that, really, this was perhaps taking it all a little too seriously. "I emailed, 'Hey, you bought the game; you can do what you want with it.'"

"Never," says John Roycroft, "play Jenga with a civil engineer." As a maxim to live life by, this is admittedly somewhat specialized—it is unlikely to appear on car bumper stickers anytime soon. But in the world of block-based tower games it contains an undeniable truth.

John, from Britain's Institution of Civil Engineers, thinks Jenga is an excellent game. "It is based on such a simple principle, yet it is a fantastic tool for looking at the reality of load paths and gravity," he says, in a slogan the game's marketing department inexplicably overlooked. "There are certainly many lessons and realities that real construction and Jenga share." This is not the only reason he thinks it's an excellent game, though; he also happens to be rather good at it.

"I recently challenged an architectural colleague; as an engineer I instinctively removed the blocks from near the top, whilst my opponent insisted on removing the blocks nearest the base." Engineers, to be clear, view architecture in the same way as football players view soccer.

It is not just by understanding the importance of good foundations that John argues we can all best defeat architects. There are also more subtle techniques. "Friction increases with load—remember this can work to your advantage but also your disadvantage," he says.

Blocks at the bottom should in theory be harder to remove, because of the weight resting on them, which increases the lateral friction. The game is designed so that all blocks are slightly different, though, so this might not be the case—a skinny block at the bottom could be loose while a fat one at the top could be wedged. This is when, applying the principles of the lever, you should be cautious. "If a block at the top doesn't slide out easily, leave it—a small nudge or horizontal lateral load at the top will have a far greater effect than the equivalent load applied at the base."

Ultimately, the crucial constraint to be considered is not friction, though, but something even more fundamental. "Trust gravity," he says.

"Always trace in your mind the load path from the top to the bottom of the tower."

A normal tower will fall over if its center of gravity is overhanging its base. In Jenga, as no one row is attached, the situation is even worse. It is less like a firmly constructed skyscraper, more like a tottering ruin. Because the Jenga tower is free to move at each level, it can be thought of as a series of nested towers—from the bottom to the top, from the second last row to the top, from the third last row to the top, and so on. If any one of these has a center of gravity that threatens to overhang its bottom row, you're in trouble—or, if you've just had your go, your opponent is. Especially if they're an architect.

FACT

Three billion Jenga blocks have been sold. If they were all used to build one tower with the same height-to-width ratio as a standard Jenga tower, it would be 100 meters (328.08 feet) high and 30 meters (98.42 feet) wide—a little bit taller than Big Ben, but a lot wider. If those blocks were instead used to play a perfect game—one in which there was a single block on each row—you could build a tower 100 times higher than the International Space Station.

ROCK, PAPER, SCISSORS

WHAT'S IT ALL ABOUT?

Decision making, for people without coins to toss.

HOW DO YOU PLAY?

Scissors beats paper, paper beats stone, stone beats scissors—this is the game equivalent of an ouroboros. Or, for those less classically minded, of a dog on YouTube chasing its tail. On the count of three, two people offer up a fist (stone), a palm (paper), or a horizontal V sign (scissors) and, after a set number of rounds, you see who has the highest score.

HOW DOES IT END?

Normally with the loser saying, "Best of seven?"

ANALYSIS

The analysis could not be clearer. For a group of rational players, no strategy gives an advantage at rock, paper, scissors. Luckily, there is no such thing as a rational group of players. Even more luckily, most players are, in fact, predictably irrational.

In 2014 a team of Chinese scientists (and, one presumes, a phenomenally indulgent grant-awarding body) paid 360 students each to play 300 games of rock, paper, scissors. What they were looking for was patterns—unconscious tics common to many players that a player in the know could exploit. Their conclusion?

"We found that the complex competition behaviors in the repeated rock, paper, scissors can be understood by a rather simple decision-making mechanism of conditional response, namely, the outcome of the last play is most influential to the next choice," says Dr. Hai-Jun Zhou, who generally plays rock, paper, scissors to decide "who should wash the dishes or clean the room."

It turns out that initially people choose scissors, paper, or stone with a probability of around one-third, distributing their affections randomly. By round two, however, it seems impossible for people to escape the clutches of irrational human psychology. Players who have just won are more likely to stick with their previous choice—after all, it worked once, so it might do so again. For exactly the same reason, losing players tend to change—and to change in a more "powerful" direction. If, for instance, they lost by playing paper, they will shift to scissors rather than stone, as scissors beats paper but stone does not.

For the serious games strategist, the conclusion is simple: If you lose a round, in the next round make the choice that would have meant you had won instead—your opponent is likely to stick with their stance. If, conversely, you win a round, look at the choice they lost on and go for whatever would lose to it: if they lost on stone, choose scissors; if they lost on scissors, choose paper; if they lost on paper, choose stone.

All of which leaves one question, how did Dr. Zhou justify to his academic superiors the request to fund the largest rock, paper, scissors contest in history? Well, the physicist explains, the game was simply a proxy. Really, he wanted to see "to what extent the approaches of statistical physics can be used to understand seemingly very complex human behaviors. We thought decision making in repeated simple game systems might be a good starting point." Really? Was that really all it was?

"Another reason, for myself, is just for fun. It's really great to go beyond the traditional subjects of physics and touch subjects that have free will and can adapt their behaviors constantly." Also, presumably, it means he has to wash the dishes rather less often.

FACT

If in doubt, go with paper. According to the World Rock, Paper, Scissors Society (yes, it exists), in competition play (yes, it exists), scissors are used 29.6 percent of the time—just under a third, and enough that using paper gives you an edge.

CONNECT 4

WHAT'S IT ALL ABOUT?

Originally marketed as "vertical checkers," which is about as useful as referring to walking as "horizontal rock climbing." Luckily, its name, Connect 4, is all the description required—just connect four tiles to win.

HOW DO YOU PLAY?

Drop tiles into a tower grid until there are four in a row.

HOW DOES IT END?

Ideally, with good players, in *Zugzwang*—the German term for being compelled to move when every move is bad. One feels that, as with Eskimos and snow, the fact that the Germans felt it necessary to define such a situation tells you a lot about the general historical milieu of Central European politics.

ANALYSIS

It was 2009 and Kanye West was in Vegas. Beyoncé was there, too, visiting for the grand opening of her husband Jay-Z's new club. So what did the trio, surely collectively the most powerful gathering in music at the time, do to celebrate? They had a Connect 4 tournament, of course. Even in hip-hop, the preeminent two-player strategy game of the 1970s has a big following.

Beyoncé won the first match against Kanye West, and the second. She trounced him in the third and humiliated him in the fourth. By the time the man who once said, "I am the greatest. I am God's vessel" had lost his sixth match, things were getting embarrassing.

Jay-Z, who prefers in his music to describe women as "bitches" and "hos" rather than intellectual behemoths, wisely decided to sit things out. The eventual score? Nine to Beyoncé, one to West.

What had made the difference? What had happened to God's vessel? Well, so the rumor in the showbiz press had it, Beyoncé had been reading the master's thesis of Victor Allis. "For a week," Victor tells me of when his daughters read the news, "I was the coolest dad in the world."

It is safe to say that for Victor, a computer scientist from the Netherlands, this was an unusual feeling. In 1988 Victor solved Connect 4. Through a mixture of strategy and computing power he proved that, with best play, whoever went first would always (just) be able to win. In doing so, he also outlined a set of tactics that ensured that even those going second would have a good chance of beating anyone. Even God's vessel, Kanye West. The most important of those tactics is to recognize what Victor calls odd and even threats.

"I have always liked to play the game," Victor says. "What I realized, before writing the program, is that I won a lot. When I thought about why I won, it was the odd and even concept—it is the fundamental rule underlying the game. It means at a very early stage you can know whether you are going to win."

A Connect 4 game between two closely matched players often ends the same way. Both players get a few sets of three tiles in a row. These sit on the board, their end dangling threateningly in midair, waiting for the other player to put their tile below, so they can be completed. The rest of the board then fills up, leaving free the column containing this threat until, eventually, the players have to fill the column and—depending on whose go it is—one player is forced to put a tile where she doesn't want to, giving victory to the opponent.

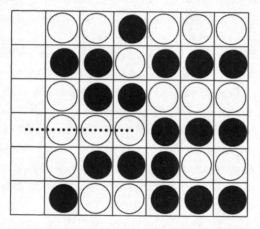

Whichever player is eventually forced to do this often feels like they have lost through sheer bad luck. "When I play other people they get to the end and say, 'Ah, I'm so unlucky.' No," says Victor, who now runs the software company Quintiq, "they are not unlucky."

In fact, Victor's entire master's thesis was about showing just that. To get the full understanding, if you are the sort of person who attains full understanding from reading 60 pages of degree-level mathematics, follow Beyoncé's lead and download the PDF. For everyone else, here is the core concept.

There are 42 spaces on a Connect 4 board—seven columns by six rows. If six columns are filled, that means 36 tiles have been placed. Half are red, half are yellow. The 37th tile, the one that has to go at the bottom of the unfilled column, must therefore be filled by the player who went first. If the other player has their threat on the second row of that column, this is bad news for the first player, who is forced to facilitate their victory. Conversely, if the first player has a possibility to complete four tiles on the third row of that column, it is very good news indeed.

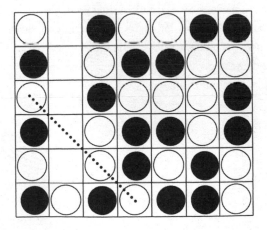

This scenario happens all the time, and normally holds true even if a column does not remain empty.

There is a very simple way to exploit it and make sure it is always your opponent who is forced to make the move they don't want to. If you go first, ensure your threatened four in a row will be completed by putting a tile on an odd row—the third or fifth (the first is no use, because your opponent will block a threat there instantly). If you go second, ensure it is completed by putting a tile on an even row—the second, fourth, or sixth. And, whatever you do, never challenge Beyoncé.

FACT

While, with best play, the first player always wins in Connect 4, that is only if he or she takes the center slot. If the first player chooses the slots on either side of the middle, then it's a draw, and with any other slot, the second player should win.

BATTLESHIP

WHAT'S IT ALL ABOUT?

At the Battle of Copenhagen, Admiral Nelson famously disobeyed a semaphored order from another ship to retreat—holding a telescope to his blind eye and saying, "I see no ships."* Eventually someone liked the idea of a naval battle in which the admirals don't get to see ships so much that they based a game on it.

HOW DO YOU PLAY?

Fire blindly at your opponent and hope you hit something. Wonder why, as admiral of a fleet that includes an aircraft carrier, you can't send some planes off as artillery spotters.

HOW DOES IT END?

As any scholar of 1970s advertising classics knows, with the exclamation, "You sank my battleship!"

ANALYSIS

HMS *Dreadnought* was formidable. Launched in 1906, she was the first ship in the British Royal Navy to be armed only with big guns—a bristling array of  twelve-inch cannons that outclassed and outranged every other battleship currently at sea. What is more, her modern steam turbines provided such propulsion that even those sensibly

---

*Actually, he said the rather less dramatic "I see no signals." Don't worry, though, his other aphorisms are intact, including "Desperate affairs require desperate measures," and one that echoes down the generations, "Hate a Frenchman as you hate the devil."

inclined to turn and flee had that option removed. *The Times* of London described her launch as instantly rendering all those other warships obsolete. It was not hyperbole—like a geological epoch, battleships would come to be referred to as pre- or post-*Dreadnought*. All boats that came to be modeled on her design were called, generically, dreadnoughts, and their construction was an arms race that, by some accounts, precipitated World War I.

But what was less appreciated at the time was an arguably even more significant fact: the extremities of her eventual artillery range, over ten nautical miles, meant that she was able to fire over the horizon at targets not visible from the deck. The consequence of this? The world had at last entered a new age—an age in which a board game based around randomly lobbing shells at an opponent whose ships are hidden from you made sense.

Admittedly, it doesn't completely make sense. Battleship is, after all, a game in which your naval opponent also finds him- or herself obliged to report back on the effectiveness of your targeting—a courtesy enemy fleets don't normally afford each other. It is also a game in which, in what some would consider a modest oversimplification of naval warfare, no ships are able to sail anywhere.

So it would be wrong to say that the tactics of naval gunnery can be taken over wholesale into Battleship, but there are certainly some basic principles, such as, be wary of putting your ships on the edge of the world and, when you've hit an enemy once, keep going until he's sunk. And, um, that's it.

For the rest, we need to go to computer scientists. In 2009, on the programming website StackExchange, a challenge was issued: there was to be a Battleship coding competition to see who could make the greatest AI admiral. Programs were

invited for submission, the competing fleets sailing into battle
thousands of times against each other.

Around the world, an arms race began. First to respond
were the programming equivalents of Viking longboats. Sturdy,
reliable, effective against unarmed monks, they did the job of
annihilating poorly organized opposition without being
spectacular. Most employed similar tactics: place your own
ships randomly on the board, then shoot randomly. When
you score a hit, check the squares around and keep going until
your opponent is sunk.

Next, Battleship technology, figuratively speaking, entered
the galleon age. They realized that, given that their longboat
opponents were employing an attack strategy in which they
searched all the surrounding squares after registering a hit, it
would be better not to be on the edge of the board—where there
is one less square to search. They stopped placing their own
fleet randomly.

Soon, it was time for—to extend the nautical metaphor—
the programming ironclads and a major improvement in
artillery accuracy. If placing your own ships randomly is a bad
idea, does the same not go for shooting? The smallest vessel in
the Battleship fleet, at two squares long, is a patrol boat. This
means that if you hit every other square, alternating by row, you
are guaranteed to place at least one shot on every ship. With this
one innovation you have halved the number of potential squares
you have to target.

Then, steaming over the horizon, firing a shot across its
opponents' bows, came the program that would render them all
obsolete. Its name? Dreadnought.

Its creator was Keith Randall, now a software engineer at

Google. What was his special insight? Is he, perhaps, a keen scholar of naval strategy and history? Or just a dedicated player of Battleship? Neither, really. "I'm not a particular fan of the game. I just saw the competition and entered." Dreadnought, he said, is about "performing very good accounting and probability analysis." Ah, the romance of the sea.

Where his program differed from most of those that came before was that it had fewer hard-and-fast tactics—and it adapted to its opponents. "There are a couple of rules of thumb. Generally you don't want to put ships touching each other. If they are, then when they hit your ship, their sinking strategy might mean they also end up sinking the second one in the process of getting the first." By checking the squares all around, they might end up hitting two ships.

"However, if you never put ships touching, your opponent can learn that—and know there is no need to shoot in the space surrounding a sunk ship. They can take advantage." So just occasionally, his fleet did touch—and, for exactly the same reason, just occasionally he did have a ship on the edge.

When attacking, Dreadnought calculates the chances that each square could contain a ship. "It looks at where the shots are, and creates a probability map." This is something that obviously is easier for a computer than for you at the back of the classroom, hiding behind a textbook with your friend Jack, especially as it adds in data about where ships have been placed in previous games. But that does not mean that humans cannot learn from the principles. If, for instance, there is a row in which the maximum space without hits is three spaces, instantly you can say any ship of four spaces or more cannot fit there horizontally—so even if it could fit vertically, the chances that it would be on any particular square in that row are roughly halved.

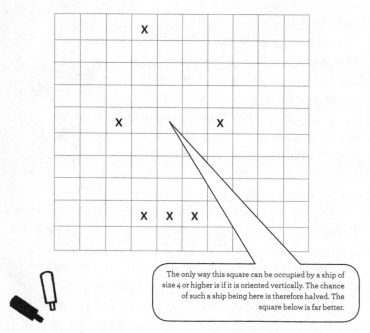

The only way this square can be occupied by a ship of size 4 or higher is if it is oriented vertically. The chance of such a ship being here is therefore halved. The square below is far better.

With each turn, think to yourself: how many of the remaining ships can fit there, and in how many ways could each of them be placed to cover that square? In that way, you maximize the efficiency of every shot.

Even so, victory is never certain: chance still plays a big role in Battleship. Keith's Dreadnought might rule the programming waves, but Keith himself does not maintain naval supremacy in his own house. "I have kids, we play Battleship," he says. "There's enough randomness in the game that you have to play lots for your advantage to show. In any particular game with my son I would say I have a 60 percent chance of winning. Also," he admits, "sometimes I let him win."

So maybe it is quite a close simulation of the real thing after all. As Nelson warned before the Battle of Trafalgar, "Something must be left to chance; nothing is sure in a sea fight."

FACT

In 2012 a film based on the game, called *Battleship*, was released. It was generally considered an extremely faithful adaptation—the board game has no plot, and neither did the film.

HANGMAN

WHAT'S IT ALL ABOUT?

Literacy education for children—using the wholesome medium of capital punishment.

HOW DO YOU PLAY?

Guess your opponent's word, letter by letter. Has he or she been deliberately annoying and chosen a word with no vowels? Or is it a double bluff, and the word does have vowels? Devious s$£*s.

HOW DOES IT END?

With a stick man swinging from the gallows, as the losing player insists he still needs a face, ears, and shoes to be drawn on. And maybe an umbrella.

ANALYSIS

If you want to send a half-decent hangman opponent to the gallows, you could do worse than choose the word *jazz*. Mind you, that assumes you can actually find a half-decent opponent. Nick Berry believes they are rarer than you think. "There's no easy way to say it," he says. "You've probably been playing hangman wrong your entire life."

Fairly early on in most people's hangman careers, they come to the realization that beating the executioner is fundamentally a matter of letter frequency analysis. Which letter can you guess that has the most chance of being in a randomly chosen word?

"If you get a couple of letters, it is really useful, and you can greatly reduce the solution set—the number of possible words it could be," says Nick, who works as a data scientist for Facebook, and so is used to terms like *solution set*. "The key thing is, how quickly can you do that?"

One common approach is for people to try all the vowels first or, gaining sophistication, look at the number of times letters pop up in the dictionary, going with the ordering ETAOIN, representing letters' occurrence in decreasing frequency. Well, for several reasons they are not as sophisticated as they believe they are.

"People think they're being smart because they read somewhere that that is the frequency of letters in the English language," says Nick. The problem is, a dictionary of likely hangman words is different from a dictionary of all words. "We have so many glue words like *the* and *on* and *an*, that it skews the distribution." If we exclude those—and if you have a friend who actually chooses *and* in hangman, you might not just need to reconsider playing hangman with them but also your friendship in general—then you get a different frequency: ESIARN.

That is just the beginning, though, for the smart hangman player.

Letter frequency analysis is a very well-developed science, in part because of cryptography, where simple letter substitution codes have been cracked for centuries by looking at which letters occur most often in written text. In some codes—and in hangman—you have more information than just the letters, you also have the length of the word. And, as an analysis by Nick has shown, while E might be the most common letter in the English language, it is not the most common letter in five-letter words.

S is. Neither is it the most common in four-letter words. That honor goes to A.

Still this did not satisfy Nick, because your failures tell you a lot, too. "Take a six-letter word," he says. "E might be the most common letter in six-letter words, and S the second most common, but what if you guess E and E is not in it?" It turns out that in six-letter words without an E, S is no longer the next best letter to try. It is A. Using this method Nick has created an attack strategy (see the table, right) that gives the best chance of guessing the first letter in any randomly chosen word.

What if you aren't guessing, though? What if you are the one choosing the word? Well, the best way to defeat someone applying Nick's tactics is to choose the word *jazz*. He has used a computer to run the entire dictionary through his system and found that that would be the last word to be guessed by anyone using his methodology.

Mind you, if you both know you are both playing by the best strategy, your opponent might want to guess *jazz* straight away. And, suddenly, Nick's strategy is no longer the best strategy at all.

Number of letters	Optimal calling order
1	AI
2	AOEIUMBH
3	AEOIUYHBCK
4	AEOIUYSBF
5	SEAOIUYH
6	EAIOUSY
7	EAIOUS
8	EAIOU
9	EAIOU
10	EAIOU
11	EAIOD
12	EAIOF
13	IEOA
14	IEO
15	IEA
16	IEH
17	IER
18	IEA
19	IEA
20	IE

FACT

Hangman probably began life as a Victorian game called Birds, Beasts, and Fishes, in which children guessed the letters in a mystery animal's name. It is not clear at what point someone took this sweet childhood game and decided to limit the guesses through a mechanism pegged to public execution.

BLACKJACK

WHAT'S IT ALL ABOUT?

More interesting than roulette, easier than craps, but unlike either, it can be (to the surprise, and annoyance, of casinos) a game of skill.

HOW DO YOU PLAY?

Keep taking cards until their value is as close to 21 as you dare. Go over, and all is lost.

HOW DOES IT END?

There are three options. 1. You win, leave the table comfortably up, and reward yourself with a nice meal and a bottle of wine. 2. You reach your predetermined maximum losses, leave the table a little wiser, and have a slightly less nice meal. 3. You discover to your surprise it is 4:00 a.m. and you are a significant proportion of your mortgage down. That's fine, though, because you've developed a super-special secret system and the next hand is definitely going to be the one that will turn things around.

ANALYSIS

After a gap of a few years, Mike Aponte recently returned to the Cosmopolitan Casino in Las Vegas. The Cosmopolitan is one of those places Vegas does so well—or badly, depending on your view. It might help to view it less as a casino than as an entire cruise ship beached in the desert.

It has three swimming pools, a shopping mall, half a dozen

restaurants, and, naturally, "canine amenities." Aponte was on his way to one of those restaurants for a business meeting, avoiding the glitzy gambling floor entirely, when he received a call on his mobile phone.

"It was Jeff Murphy." Murphy is head of surveillance at Cosmopolitan. "I answered and said, 'How the heck did you know I was here?'" Already, he would soon learn, in Murphy's office they had taken a screengrab of the CCTV footage of his arrival, printed it out, and written "Public Enemy Number One" beneath it.

Murphy knew it was Mike for the same reason that every casino in the US knows when Mike walks in. Mike is probably  the most famous card counter in the world, and if casino surveillance guards—the shadowy security teams hiding behind their CCTV cameras and two-way mirrors—have any purpose at all, it is to make sure that he never again gets to sit at their blackjack tables. That does not stop him from training other people, though.

"It's the only casino game you play against the house which has a true element of skill," he says. "The reason why, is that what happened in the past has an impact on current probability." In a typical blackjack game, the dealer uses six packs—312 cards—shuffled together. This means the probability of getting a king, say, may start off as 24 in 312, but if half the cards are gone and no kings have emerged, that probability doubles. "Ninety-nine percent of blackjack players have no idea what the odds are at any point in time," says Mike. "Card counting is a means of knowing where you stand at all times."

These days Mike provides blackjack tuition. In the

past, though, he was head of the Massachusetts Institute of Technology blackjack team. For two glorious decades in the 1980s and '90s, this team of mathematicians and scientists took most of the casinos in America for all they were worth. Their secret? "We were not gamblers."

Before joining the MIT team, Mike had never played blackjack. "Why would I gamble? Why play a losing proposition? Gamblers are not rational." His only understanding of card counting came from the film *Rain Man*, in which an autistic man wins in Vegas by recalling all previously played cards. "I thought you had to have a photographic memory." He would learn it was far simpler.

Before that, though, the first thing members of the MIT blackjack team were taught was not how to count cards, but what is known as "basic strategy" (see diagram on page 60). This is the set of optimal rules for how to play in whatever situation presents itself. It tells players when they should stick with the cards they have, fold to get another, or—a more advanced technique—split equal cards into two hands.

"The reason basic strategy is so important is that it reduces the house edge compared to the average player by about 75 to 80 percent." The house edge is the expected casino winnings for each hand. In blackjack a crucial factor of the game is that the dealer only plays after everyone else has been dealt all their cards. If the player goes bust, he or she loses—even if the dealer also subsequently goes bust. "That means that for most players, including professional card counters, you will never win more hands than the dealer."

Learning basic strategy means that for a typical player the house edge drops from 2 percent to 0.5 percent. With the addition of card counting, in blackjack—uniquely among casino games—the advantage goes the other way. The dealer might still win more games, but card counting tells you which ones those are likely to be, enabling you to bet big on the others.

Unlike the scene depicted in *Rain Man*, the technique does not involve memorizing each card; it simply involves keeping a rough idea of how many high cards remain. The more high cards there are as a proportion in the pack, the greater the chance of getting blackjack—an ace and a high card—which pays out extra. Since the dealer is obligated to fold with any hand below 17, more high cards left also mean that he or she is more likely to go bust.

Strategies for card counting vary (see the following page), but typically the player keeps a running total, subtracting 1 for every high card dealt, adding 1 for every low card, and keeping the total the same for a 7, 8, or 9. When the total exceeds a certain number—10, say—that implies a lot of high cards are left and the house no longer has the edge. That is when Mike Aponte would bet high. Until, of course, he got banned from every casino in the US.

BASIC STRATEGY

Dealer's Up-Card

	2	3	4	5	6	7	8	9	10	A
5-8	H	H	H	H	H	H	H	H	H	H
9	H	D	D	D	D	H	H	H	H	H
10	D	D	D	D	D	D	D	D	H	H
11	D	D	D	D	D	D	D	D	D	H
12	H	H	S	S	S	H	H	H	H	H
13	S	S	S	S	S	H	H	H	H	H
14	S	S	S	S	S	H	H	H	H	H
15	S	S	S	S	S	H	H	H	SR	H
16	S	S	S	S	S	H	H	SR	SR	SR
A2	H	H	H	D	D	H	H	H	H	H
A3	H	H	H	D	D	H	H	H	H	H
A4	H	H	D	D	D	H	H	H	H	H
A5	H	H	D	D	D	H	H	H	H	H
A6	H	D	D	D	D	H	H	H	H	H
A7	S	D/S	D/S	D/S	D/S	S	S	H	H	H
2-2	P	P	P	P	P	P	H	H	H	H
3-3	P	P	P	P	P	P	H	H	H	H
4-4	H	H	H	P	P	H	H	H	H	H
6-6	P	P	P	P	P	H	H	H	H	H
7-7	P	P	P	P	P	P	H	H	H	H
8-8	P	P	P	P	P	P	P	P	P	P
9-9	P	P	P	P	P	S	P	P	S	S

Player's Hand

Additional Basic Strategy

1. Never take insurance
2. Always stand hard 17 and higher
3. Always stand A8, A9, A10, and 10-10
4. Always play 5-5 as a 10
5. Always split Aces

Key

H   Hit
S   Stand
D   Double down if allowed, or else hit
P   Split
SR   Surrender if allowed, or else stand
D/S Double down if allowed, or else stand

CARD COUNTING

For every card dealt that is of value 2 to 6, **add 1**.

For every card of value 10 or a picture card, **subtract 1**.

The resulting total, which you add and subtract from throughout a game, is called the **running count**.

The importance of the **running count** value depends on how many cards are left to be dealt. Therefore, you can divide the **running count** by an estimate of the total number of decks left to give the **true count**. A high **true count** indicates you are more likely to receive cards of value 10. This means the dealer, who sticks only on 17 or above—whereas you can choose to stick below that—is more likely to go bust. It also means the player is more likely to get blackjack (ace and 10), which pays out at a higher return.

FACT

The Barona Casino in San Diego hosts a Blackjack Hall of Fame for some of the most successful players in history. Inductees are promised a free room and board for life on one condition: they must agree never to play blackjack in the casino again.

SCRABBLE

WHAT'S IT ALL ABOUT?

Do you know your zo from your za, your yo from your ya? Then you are already ahead of the casual Scrabble player, who might call "fy!" or utter an "od" at their bo's use of esoteric two-letter words.

HOW DO YOU PLAY?

Scrabble is about words in the same way that stamp collecting is about delivering mail—in other words, it isn't. Forget about making pretty words, start looking for ways to get a Q on a triple letter score.

HOW DOES IT END?

With a polite impasse (11 points, plus 50 for using all seven tiles) about whether or not *od* is a word, precipitating a somewhat socially strained visit to the bookshelf for a dictionary, which is then slammed shut with a "Fine, tell me what it means then." The answer, not that it matters, is: "a hypothetical and omnipresent natural force."

ANALYSIS

There are two answers that Brett Smitheram gives when he is asked how he got into Scrabble. The first is that it was because of the "love of language." The second is that the Under 16 national champion at the time "was my nemesis. I really hated him. I thought that the best way to get at him was to be better at the thing he was best at, and really piss him off."

There can be few better illustrations of the sort of strategic thinking and dedication that has made Brett one of the finest Scrabble players in the world—or of the contradiction at the heart of the game. For some, Scrabble is a pleasant word game, a gently taxing way to end an evening with friends and family. For others, it is a game of tactics and mathematics in which words are simply a tool—a tool that can be applied without any knowledge of their actual meaning.

"When you play Scrabble with linguists they get caught up in the loveliness of the words," says Brett. As far as he is concerned, this is like enjoying the aesthetic sweep of a Samurai sword when your opponent has an Uzi. "You might be able to play, say, *disjaskit* and have a 5 percent chance of winning, or *kid* and have a 100 percent chance. I would always play *kid*, but I've seen it time and time again, people play the nice word and get stuck and don't analyze the strategy." For Brett it is not just about the word, or even the score he can get from the word, but what effect playing that word will have on the board, on his opponent's next available moves and on his next available moves.

Before you can learn all that, though, you need the basics. "Everyone starts by learning the two-letter words." These enable "parallel play," meaning that you can have a word going horizontally that also adjoins several letters vertically along its length to make a whole series of words. The next step, says Brett, is to notice how the board works. "Premium squares are generally four or five places apart. So make sure you are stronger in four- or five-letter words, then you can reach from a triple-letter score, say, to a double word." He trains by using computer programs that generate words in the order of their probability of appearing in Scrabble tiles, and learning them.

"My third piece of advice would be to understand the relative value of tiles. Blanks are not worth anything, but are the most prized tiles in the set. You shouldn't really play a blank for less than 50 points. Similarly, S is only worth one point, but simply by using it for pluralization you maximize the chances of a seven-letter word." If those tiles are underappreciated, with others the reverse is true.

"Q is a classic example. A lot of players will hold on to it waiting for a U—thinking it is worth lots of points. The problem is, when you are finally able to play it the opponent has been scoring 30 or 40 points every go and you've been hampered by only having six usable tiles. Usually, unless there's something on the immediate horizon, it's best to get rid of it." In most cases, by "getting rid of it" he means just playing it alongside, say, an "i" to spell "qi" for 11 points. Sometimes it is worth actually losing a go to change tiles, though.

"W and U—those two tiles together are appalling. The worst three-tile combination is v, u, w." Although those three are so bad together that every serious player has learned the strategies to use them. "*Vrouw* (used in Afrikaans instead of *Mrs.*, since you ask) can get rid of them all at once."

To demonstrate the power of these simple techniques, Brett was once challenged to play a game in French, with a native speaker whose vocabulary vastly exceeded his own. Importantly, though, Brett ensured that his vocabulary in very specific areas was superior.

"Beforehand I learned the French two-letter words, and the highest-probability seven-letter word—which also makes four anagrams." This word is *aligote*. "I spent the whole game playing two-letter words and waiting to play this word."

He won, naturally, and afterward his enraged opponent stole the book he was using to practice. Such reactions are common. He doesn't play against non-tournament-level players these days, because they "become too despondent." When he plays online, he says people often call him names and accuse him of cheating. "I always say, if you go out for a race with Mo Farah you wouldn't expect to win. So why do you with me?

"People lack an awareness of their capability when it comes to language; people think they are a lot better at it than they are. This causes a lot of rage, not just online but around the kitchen table. I've lost count of the amount of times I've heard people have fallen out with their granny because of a dispute over whether something is a word."

And what of his nemesis, how did it go with him? Was there a rapprochement? "He beat me the first time, then I won all the games after that and he gave up Scrabble. It was a victory in every possible sense." Among Scrabble players, that counts as a heartwarming tale.

FACT

The muzjiks, or Russian peasantry, worked in *sovkhoz*—state-owned farms, also known as *kolkhoz*. Collectively (and they knew a lot about collectivization), these are three of the highest-scoring seven-letter words learned by serious players around the world. So it is that the blood of the workers oils the machinery of Scrabble.

See a list of two-letter words in Appendix 2 at the back of this book.

THUMB
WRESTLING

WHAT'S IT ALL ABOUT?

Float like a butterfly, sting like a bee; the thumb can't hook if the knuckle stays free.

HOW DO YOU PLAY?

First, lock fingers. Then position your thumb in front of your opponent's and let the two thumbs face off, impassively, like Mayweather and Pacquiao squaring up at a weigh-in. Would that pair be man enough for what follows though?

HOW DOES IT END?

While boxing is a rigidly coded contest, where the victor has to leave his opponent alone when he hits the mat, thumb wrestling has no rules. It is more like a dirty street brawl. If, admittedly, both participants in the brawl have no arms, one leg, and can only hop around vigorously and then lie on top of each other.

ANALYSIS

When grizzled veterans get together to compare war stories and reminisce about the old times, one fight is still spoken of. A clash of titans, an epic struggle between dogged technique and brute power. Think David versus Goliath, but with bigger stakes. It was 2001, and in contention was nothing less than the annual San Francisco thumb wrestling championships.

"I only exaggerate a little," says Oscar Villalon, recalling what has become his own Rumble in the Jungle; a contest that would echo in eternity. "My opponent was maybe nine feet tall. His hand completely enveloped mine. He was blond and a huge human being—he looked like he sailed with Erik the Red."

Ordinarily, Oscar admits he would have thought it was surely the end. But, as happens with all the greatest champions, in that moment something inspired him and gave him an inner force to try and to believe he could overcome. Specifically, in this case, that "something" was alcohol. "I'd had a few beers."

"I went for him a bit like Mike Tyson. While he was whopping his baseball bat–sized thumbs around I was ducking and bobbing, weaving on the inside. You go for where the web of thumb begins, try to pin him there, from that knuckle back there." What you don't go for is the nail—you slip right off. "The nail is nature's linoleum."

Oscar is a three-time thumb wrestling champion. Now retired, these days he is a grandee of the sport rather than a competitor. "It's a young man's sport. By the end of a tournament your forearm is pretty numb," he says. He realized that his third title could well be his last. "After three wins I just wanted to go out on top; the last contest was brutal, absolutely brutal."

Even if his flesh is weak, however, his tactician's brain is still as strong. Oscar has a theory: he believes thumb wrestling has a lot in common with another sport of pugilism.

"I don't mean in any way to diminish the sport by making this comparison, but it's pretty much about employing the tactics of boxing," he says. The sport he is concerned about diminishing through this comparison, by the by, is thumb wrestling, not boxing.

In one respect, though, he maintains they are the same. "You have to take leather to give leather."

"The crux of thumb wrestling is in the counterpunch—the counterthumb. You need to set up your opponent so that he or she keeps stretching their thumb out to get you." It is when their thumb is extended, unguarded, that you strike. "If you are quicker than they are, then you get in over the top and pin."

Equally, just as a good boxer doesn't look at the opponent's gloves, so the same principle applies in thumb wrestling. "It's not about the thumb. Do. Not. Look. At. The. Thumb," he says, enunciating each word. "One of the things I do to unnerve my opponents is I look at the ceiling when I thumb wrestle. Just as in boxing, you do it by feel—you feel them out with your jabs. If you look at the thumb you will lose."

Eventually, though, you have to accept that tactics will only get you so far. For Oscar, his most fearsome opponent was arguably not the nine-foot Viking, but someone with an even more insidious advantage.

"Just occasionally you get someone with inordinately sweaty hands, and you just can't get a purchase on their skin," he says. "It is the thumb wrestling equivalent of fighting a southpaw. They have ample opportunity to get you because they can come over the top really easily—they are slick with sweat. And your hands are normal because you're a normal human being.

"Personally, I think if the sport goes big they should regulate that sort of thing. Your opponent should be forced to wear a latex glove." While that may be the goal, and we can all agree he makes a good case, even now there is yet to be a proper regulating body for thumb wrestling. So Oscar had no choice but to fight.

"It was really, really hard. It was an ugly fight. I just had to rely on technique and cussedness." And, he admits, an occasional bending of the rules about how much you're allowed to bend your wrist. "There was maybe a little tilting of the wrist on occasion. But if the referee didn't see it, it didn't happen."

Ultimately, it was the same dogged determination that won the day against his blond giant. After fifteen seconds, the fight was over. The Viking was vanquished—brain had triumphed over brawn. Either that, or—Oscar concedes that it is an unlikely but just possible alternative—"he just didn't care that much about thumb wrestling."

FACT

According to palmistry a long, flexible thumb—the kind arguably ideal for thumb wrestling—is also associated with a logical mind and keen intellect. Which, ironically, usually means possessors of such thumbs don't believe in palmistry.

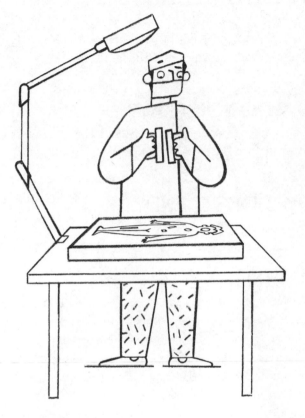

OPERATION

WHAT'S IT ALL ABOUT?

Cavity Sam is ill. So ill, in fact, that he needs all his organs taken out slowly, one by one. This makes him better until, eventually, he dies. Good family fun.

HOW DO YOU PLAY?

A knowledge of anatomy sort of helps, a shaky hand definitely hinders. Just like real surgery.

HOW DOES IT END?

Badly, if you're Sam.

ANALYSIS

When Roger Kneebone, Professor of Surgical Education at Imperial College, sees a good surgeon in action, he says it is "like a miniature ballet.... There's a kind of choreographic beauty to it. The most impressive surgeons are the ones with the most fluidity, who just slowly, efficiently make single movements with no apparent rush."

As with lifesaving operations, so with board-game operations. "Success in Operation—whether the board game or the real thing—is about the relation between the instrument, the hand, and the job," says Professor Kneebone—and that really is his name. Surgeons long ago learned that steadiness requires different grips from those we would more naturally use in our daily lives. "For really delicate work, don't hold the tool like scissors. It helps a lot to grip with your thumb and ring finger, then use your index finger for supporting it. If real exactness is required, you might even want to support with your other hand."

R. M. Kirk's canonical handbook, *Basic Surgical Techniques*, has other advice. "Surgeons do not usually have extraordinarily steady hands," it says, noting that what abilities they do have diminish with age. With that in mind, the handbook explains, "If you hold instruments at arm's length, the tips magnify the tremor, and anxiety exaggerates this. Do not feel embarrassed. The further the distance from a firm base to the point of action, the less steady are your hands." So, instead, it advises, "Stand upright with feet apart, arms and fingers outstretched." This is, admittedly, somewhat complicated as a technique if the Operation board is on the living room floor.

The next suggestions are potentially more pertinent. "Press your elbows into your sides and you should find your hands are steadier. Sit, or brace your hips against a fixture, to become even steadier. Rest your elbows on a table; also rest the heel of your hand or your little finger on the table." Finally, it says, "If you need to carry out a smooth movement, try practising it in the air first, as a golfer does before making a stroke." Which might make you look like a total pillock, but you'll be a pillock who wins, rather than someone with dignity who loses. And if this book is about anything, it is about the willing sacrifice of dignity on the altar of victory.

It is through a combination of such techniques, says Professor Kneebone, that some surgeons manage to keep going past an age when their hands start to shake. Any other insider tips? "It's a bad idea, most surgeons agree, to have a horrible hangover."

FACT

In 1964, John Spinello received $500 for inventing Operation. In 2014, fifty years and $40 million in sales later, the seventy-seven-year-old Spinello realized he needed a real operation—and couldn't afford it. Fans of the game crowdfunded the money for him.

RISK

WHAT'S IT ALL ABOUT?

Had human history been different—had, say, Australian aborigines developed gunpowder—this is how things would have gone: biding their time, they would have waited for Genghis to rise and fall, and Charlemagne to do his thing, then they would have struck like a whirlwind, sweeping across the steppes and up through Siberia into the US. At least, that was what happened last time you played Risk. It is, admittedly, an outcome that presumes that aborigines also developed thermal underwear and snowmobiles.

HOW DO YOU PLAY?

Fool your opponents into thinking you are weak, as they fight with laughable futility among themselves. Then, when the time is right, your strategic might will annihilate their pathetic armies. Except, perhaps, for a single soldier left shivering fearfully in Kamchatka, so you have something to toy with.

HOW DOES IT END?

With you alone, shivering in Kamchatka, wondering why you insist on playing games with such sadists.

ANALYSIS

"Victorious warriors win first and then go to war," said Sun Tzu, "while defeated warriors go to war first then seek to win." It turns out that brutal conflict in the first millennium BC has a surprising amount of overlap with a dice-based strategy game popular with computer scientists.

Risk works as a game because the probabilities—of success, or failure—are just slightly too complex for people to understand them intuitively. And so, instead we attack while uncertain of the outcome, letting the attrition of our men decide the result. A bit like Stalingrad, but with rather more hinging on getting a six, rather less hinging on getting Tractor Factory Number Six.

Just as in real war, you can never remove the element of chance—the rainstorm that left Napoleon's Waterloo guns stuck in the mud, the little scrote of a nephew who persistently rolls just the required number—but if you understand the chances, the game suddenly becomes as predictably unpredictable as snakes and ladders.

So, on the downside, it loses all excitement; on the upside, with a little bit of patience, you win. As Rommel said, "Sweat saves blood."

What, then, are the probabilities? Dice are at the core of Risk. When an attacker meets a defender, the attacker can roll three dice (assuming he or she has three armies or more), the defender two (assuming he or she has two or more). The defense advantage comes from winning if two dice are tied.

With each roll there are three outcomes. 1. Two of the attacker's dice are higher than the defender's; the defender loses two armies. 2. All of the attacker's dice are lower than, or equal to, the defender's; the attacker loses two armies. 3. Both lose one army. The question is, in any given battle, who is most likely to win?

It would be easy to answer if you were just throwing two dice each and seeing who has the highest, but this situation is a little more complicated. In fact, it is almost certainly a lot more complicated than the Risk creators ever intended, involving a branch of degree-level statistics that has been in existence for only a little more than a century.

Superficially, it seems relatively simple to work out who is most likely to win a given skirmish. Most battles that decide a game will involve the clash of big armies—in which the attacker rolls three dice against the defender's two, and you keep on rolling until one side is destroyed.

The smallest battle with equally matched armies in which you can throw this dice combination is three versus three, and the defender wins 53 percent of the time. Since any bigger clash of armies is going to involve multiples of such smaller clashes, it appears clear that the defender has the edge.

What seems clear in probability theory is not necessarily so, however. In 2003, Jason Osborne, now a professor at North Carolina State University, applied a technique known as Markov chains to look at the statistics. What he found was that the idea that defense is the best policy is an illusion. For any evenly matched conflict of more than five armies a side, the attacker has a decisive, and growing, advantage.

For Professor Osborne, the conclusion is obvious. "The chances of winning a battle are considerably more favorable for the attacker than was originally suspected. The logical recommendation is then for the attacker to be more aggressive." Of course, this is only in the aggregate. In Risk, perhaps even more than in actual war, the best guide might not be Sun Tzu but Napoleon, who said, "I have plenty of clever generals but just give me a lucky one."

Which is something to console yourself with if you still find yourself forced to defend a small patch of Siberia while your nephew sweeps across Asia like a latter-day Khan.

If both the attacker and defender are rolling the maximum number of dice, the chances of both losing an army is 33.6 percent. The chances of the attacker losing two is 29.3 percent. The chances of the defender losing two, however, is a whopping 37.1 percent. This is why, ultimately, the attacker has the advantage. The only reason why smaller battles tend to favor the defender is that they have a greater chance of ending up in a two versus two situation, the one occasion when there is a significant advantage for the defender.

Probability that the attacker will win

Defending Pieces

Attacking Pieces	1	2	3	4	5	6	7	8	9	10
1	0.417	0.106	0.027	0.007	0.002	0.000	0.000	0.000	0.000	0.000
2	0.754	0.363	0.206	0.091	0.049	0.021	0.011	0.005	0.003	0.001
3	0.916	0.656	0.470	0.315	0.206	0.134	0.084	0.054	0.033	0.021
4	0.972	0.785	0.642	0.477	0.358	0.253	0.181	0.123	0.086	0.057
5	0.990	0.890	0.769	0.638	0.506	0.397	0.297	0.224	0.162	0.118
6	0.997	0.934	0.857	0.745	0.638	0.521	0.423	0.329	0.258	0.193
7	0.999	0.967	0.910	0.834	0.736	0.640	0.536	0.446	0.357	0.287
8	1.000	0.980	0.947	0.888	0.818	0.730	0.643	0.547	0.464	0.380
9	1.000	0.990	0.967	0.930	0.873	0.808	0.726	0.646	0.558	0.480
10	1.000	0.994	0.981	0.954	0.916	0.861	0.800	0.724	0.650	0.568

There is a reason why Britain still holds on to Gibraltar, and it is not just so that we can get photographs of monkeys sitting on pillar boxes. It is the same reason why we fought a disastrous war over Suez and an even more disastrous war over Gallipoli.

These small strips of land control the access to even bigger ones—dividing Asia and Europe, Africa and Asia, the Mediterranean and the Atlantic. Just as in real life, controlling the choke points in Risk is crucial. Even more crucial is knowing where they are.

In the same way as the map of the London Underground simplified train travel to what is necessary for efficiently getting round London, so a paper by MIT student Garrett Robinson has simplified the Risk board to what is necessary for efficiently crushing your opponents. In his study, titled "The Strategy of Risk," he produced a handy diagram—in which you can quickly see, say, that with just one entry and exit point, Australia is a continental fortress, while with five, trying to hold Russia could be your undoing. Just ask Hitler.

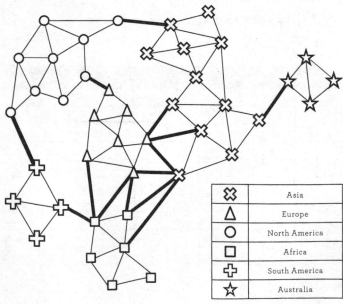

✖	Asia
△	Europe
○	North America
▢	Africa
✚	South America
☆	Australia

Bold lines show links between continents.

FACT

Among many novelty releases over the half century of Risk's existence has been Risk: *Narnia*, Risk: *Lord of the Rings*, and Risk: *Star Wars*. Unlike most themed board games, they are not just gimmicks: with a different map, you get a markedly different game. Defending Middle Earth involves considerably different tactics than defending the Middle East.

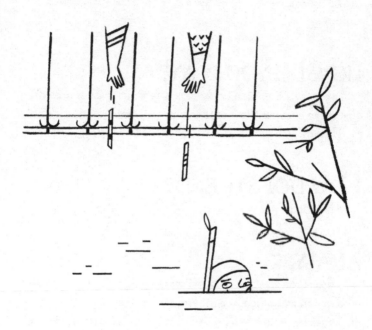

POOH STICKS

WHAT'S IT ALL ABOUT?

A whimsical attempt to evoke A. A. Milne childhoods that never were, this is a game generally instigated by parents trying to show children that wholesome walks in the great outdoors can be fun—and are viewed as proof of precisely the opposite by children.

HOW DO YOU PLAY?

Drop two sticks off a bridge and see which passes under first. Just make sure you start at the upstream end. And that it's not a road bridge....

HOW DOES IT END?

Anticlimactically.

ANALYSIS

When Professor Ian Guymer wants to illustrate the complex fluid dynamical processes inherent to his research on pollution dispersal, he has a go-to PowerPoint image: Winnie-the-Pooh. "My work is about predicting the movement of water. Where does it go, what is its path, when will it arrive somewhere?" he says.

This makes his research crucial in understanding what to do when our rivers are contaminated with effluent—one kind of poo stick. It also makes it crucial (if, admittedly, a little less important) in understanding the intricacies of the other kind of pooh sticks—the game invented, or at least immortalized, by the residents of the Hundred Acre Wood.

Water engineers actually consider pooh sticks one of the more useful tools of their profession—although they generally don't use sticks. No, they prefer fruit. "Oranges are the best. They help show you the flow, and biodegrade so environmentalists don't get angry," says Professor Guymer. They are also almost neutrally buoyant, floating just below the surface. But oranges are not the only fruit for a fluid dynamicist seeking to trace a river's path. "Depending on where you are in the world, the fruit changes. If you are on a really nice assignment, you use coconuts."

What has a lifetime of seeking assignments with coconuts taught Professor Guymer, who sits on Britain's Institution of Civil Engineers' water panel? Well, first, go for the middle. "If we look at the simplest possible shape—a rectangular concrete channel—then water velocity varies just due to distance from the boundary." Because water experiences friction when it comes into contact with the edge, it is slower at the sides and the bottom. So, in this case, drop your stick as far from the edges—including supports on the bridge—as possible.

If all rivers were uniform rectangular channels, though, Dr. Guymer probably would not have a job. In the real world they are often a lot more complex than that. "The basic problem with a natural river is that you've got spatially varying flow velocity," he says, not exactly echoing A. A. Milne. Most important, at least as far as the serious pooh sticks competitor is concerned, there are eddies generated by obstructions upstream that can cause circular flows. "If your stick gets trapped in that, it will eventually get out but it may have to do one or two rotations," he says.

His final piece of advice does not require the skills of a civil engineering PhD to formulate. "Just make sure it doesn't get stuck," he advises. Choose a small stick—the bigger it is, the more likely it is that it will hit an obstruction. Preferably, you should also select one without anything that can get caught on reeds or the bridge. Depending on where you are, you could even go for an orange instead.

FACT

A. A. Milne's Pooh books may have created an enduring game, as well as an enduring character, but not everyone was a fan of his whimsy. When Dorothy Parker reviewed *The House at Pooh Corner* for *The New Yorker*, under the nom de plume Constant Reader, she came across one passage written in Milne's characteristic style that deliberately included spelling mistakes. "And it is that word 'hummy,'" she wrote, "that marks the first place … at which Tonstant Weader Fwowed up."

SLOT CAR
RACING

WHAT'S IT ALL ABOUT?

A driving simulator for a generation without computers that largely persists in a generation with computers because it's their dads doing the buying.

HOW DO YOU PLAY?

Re-create all the thrills and spills of the golden age of Formula One—the danger and the excitement, the technical wizardry and the split-second timing—by pressing and depressing a single plastic button.

HOW DOES IT END?

When your mother/wife insists she needs the dining room table back.

ANALYSIS

Johnny Herbert is one of motor racing's great all-rounders. As a Formula One driver, he competed for eleven years for seven teams, was on the podium seven times, and won three races. Before that he won the Le Mans 24-hours, the Formula Ford Festival, and British Formula 3. And, before even that, he came third in the South of England Junior Scalextric (slot car racing) championships.

"I'm a slot car boy. I used to spend hours on it in my parents' loft," he says. Was this where his career really began? If so, what can the keen slot car aficionado learn from a keen Formula One racer? Well, he argues, at least some skills are transferable—if not the ones about continually getting little metal brushes aligned so they don't get caught on the chicane. "It is a bit different from proper racing," he says. "You only have a single trigger." Even the most ardent slot car fan would have to acknowledge that the absence of gears, braking, and, indeed, a steering wheel, are something of a limitation in a driving simulator. "But there are links. Smoothness really comes into play, and so does anticipation. Be smooth on the trigger when you're going back on the throttle, but not when braking. And keep watching: as you go into the corner, that is the time to reapply the trigger."

Just as in the real thing, he advises having practice laps. "When you know what the car is doing, that's when you will know if you have pressed slightly too much and got a slide. Learn lap by lap what the car is doing, so you can respond. Really it's just about anticipation, then squeezing the trigger slowly at the right time."

FACT

The fastest slot car was a replica of a Honda F1. It reached 31 mph or, in "scale mph"—the term slot car enthusiasts use to compare speeds in the 1:32 track to those in real life— it comfortably exceeded the sound barrier to reach 984 mph.

PAPER
AIRPLANES

WHAT'S IT ALL ABOUT?

Origami, for people with very little interest in swans.

HOW DO YOU PLAY?

Ideally, solve the Laplace equation for noncompressible two-dimensional airflow over a semirigid wing. Then, adjusting your design accordingly, wait until the teacher's back is turned and throw. The first bit is optional.

HOW DOES IT END?

With detention.

ANALYSIS

John Collins's great insight came in an aircraft hangar, with a paper plane that refused to go straight. "The plane would go halfway down the hangar, then suddenly it would make a U-turn. I was kind of at a loss—it was just doing things a plane shouldn't do." Every time he threw it, the same thing happened—the beginnings of a great distance throw then, without warning, it boomeranged.

Luckily for him, in his attempt to break the world distance record for paper airplane throwing, this hangar contained people rather used to making planes do things they shouldn't do: it was the home of Scaled Composites, the US company that also makes suborbital space planes.

"One of the guys who designs the ignition systems on their engines was watching us fly." He thought he knew what was happening. "There was an anomaly on the wing, and the air was hitting it at different places depending on the speed." As the paper plane slowed, the airflow changed. Suddenly at this point a small difference in how John had folded the wing became significant.

This gave John an idea.

Paper planes had always been part of John's life. Over the years he has come to appreciate the deep insights they give into aeronautics, and has gone from being a boy who liked to throw them to being a man who organizes paper plane contests for children to get them into science. And also, admittedly, being a man who likes to throw them.

In his attempt at the distance record he quickly came to two conclusions. First, he realized he needed to abandon the classic dart design. "We quickly maxed it out. I realized we could only get it so far—we had to switch to a glider wing." He needed something broader, with more lift. Second, he came to the sad realization that someone else would need to throw it—someone with a stronger arm. "A good throwing technique is key." His first thrower "had big hands and kept crushing the planes." His second one "had a snappy throw and cut the planes in half." The third one, Joe Ayoob, was just right.

But he still had the problem that afflicts all attempts to throw a paper airplane as far as possible: the need to compromise between lift and speed. "When you are going fast, initially you want low drag," he says. This means having the plane angled so that the nose is parallel with its travel. "When it is going slowly, you want stability." This means having the nose pointing slightly up. That day in the aircraft hangar he realized that if he could just design a plane that changed how its wing is angled in the air depending on its speed, there would be no need for compromise at all.

He began folding anew, but this time with one small change to the "dihedral angle," the angle made where the left and right sides of the wing meet. "The shape of the wings was different. They have a flatter dihedral angle at the nose, and we cranked up the dihedral angle farther back toward the tail." The result was a plane with low drag at high speeds and high stability at low speeds.

Six weeks later it flew 226 feet and won a place in *Guinness World Records*. It also ensured that paper plane design would never be the same again. You don't have to be a rocket scientist to design a good paper airplane, but knowing one certainly helps.

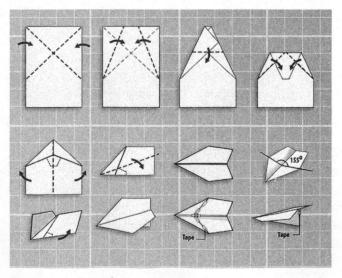

Instructions for folding the Suzanne paper airplane

Full instructions for folding this record-breaking paper plane are in John Collins's book, *The New World Champion Paper Airplane Book*, published by Random House US, and there are video guides online for John Collins and the Suzanne paper airplane that will walk you through the process of folding it.

Until recently, every time a physics teacher was hit by a paper dart when their back was turned, they should have considered it an affront. Not to their authority, but to their subject. Because for most of the time that people have been making paper airplanes, nobody has known how they fly.

In school we are taught how planes work. It begins with the airfoil—the curved cross-section of a wing. In the simplest understanding, this shape, longer on the top than the bottom, forces air to go faster along the upper edge, creating a pressure differential that generates lift. More complicated explanations might take in vortices, or the cutter condition, or Bernoulli's theorem. But it begins with the airfoil.

The problem with this is, paper airplanes don't have airfoils. Their wings are completely flat—their passage across a classroom a minor aerodynamic miracle. Curvature can have no role in how they generate lift. So what does? It turns out it all comes from the angle of the wing relative to the airflow.

"Until the 1950s we didn't really know how they flew," says Steve McParlin. Steve is a fellow of Britain's Royal Aeronautical Society, and now teaches aerodynamics. The bulk of his career, or at least the unclassified version, was spent working on combat aircraft and supersonic wings. It is in that work, on the fluid dynamics of wings designed to pass the speed of sound—whether on their way to intercept Soviets or not—that we find an unlikely congruence with paper airplanes. Both of them have to have flat wings: supersonic planes because it minimizes drag, and paper airplanes because, well, they're made out of paper.

Now we do know how they fly, and it has very little to do with airfoils. All of the lift is generated by the angle of the wing relative to the airflow. This has some unusual and less-predictable consequences.

When the air hits it at an angle, there is higher pressure on the lower half of the wing than the upper, generating a lift force. While that is an explanation for lift that almost anyone can understand,

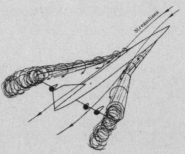

Vortices on double delta wing.

unfortunately it explains only a very small part of it. The rest comes from what happens when the air buffeting the wing goes round its edges.

"In the analytical solution, what happens is you get an infinite velocity as it goes round the corner," says Steve. This is another way of saying all his equations go wrong at the edges. "In real life," he adds comfortingly, "we don't have infinite velocities." Instead, above the wing the airflow curls up into a vortex.

In conventional aerodynamics, that is the last thing you want. "Vortex flow was always considered a source of drag. It's only when the wing is very slender that vortex flow starts to form a coherent structure, instead of something that is horribly messy." And when they do form a coherent structure, they also generate lift, sucking the wing from above.

FACT

John Collins might hold the world distance record, but his plane only stayed up for 9 seconds. The world time record is 27.6 seconds and involved throwing the plane vertically before it slowly circled down.

CHECKERS

WHAT'S IT ALL ABOUT?

"Everybody plays draughts,"* said Frank Dunne, an early twentieth-century writer on the game, "but comparatively few are draughts players." Unlike chess, you can be taught the rules in a few minutes, but the problem with this rapid learning curve is that it leads beginners to mistakenly believe they are not, to paraphrase Dunne, playing like an idiot.

HOW DO YOU PLAY?

Set up your twelve white and twelve black pieces on the black squares, with two rows free in the middle. Move a selected piece one square diagonally forward per move, or remove an opponent's piece by jumping over it when adjacent, and if the next square is free to land on. If it is possible to take a piece, you have to. When you get a piece to the other side, it can move diagonally backward, too. See? Told you the rules were simple.

HOW DOES IT END?

Always with a massacre. Chess might allow for the relatively civilized option of regime change—remove the king and it's all over—but checkers draws from a take-no-prisoners, Mongol hordes approach to war. The loser is the one who is completely destroyed.

ANALYSIS

The first thing that happened, after Jonathan Schaeffer completed the program he had been working on for almost

---

*Draughts—or drafts—is another name for checkers.

twenty years, was that the hate mail began. "They said I was going to destroy the game, to ruin it—that no one was going to play." The world's checkers players were furious: Professor Schaeffer had created a computer opponent that could never lose.

Schaeffer is, by his own admission, a bad checkers player. Luckily, though, he is an excellent computer scientist. Starting in 1989, he began working on a program that could "solve" checkers—playing the optimal strategy whatever move its opponent made. It was a daunting task.

"There are 500 billion billion positions. Five hundred billion billion is not something people really understand. So I use an analogy. Imagine you drain the Pacific Ocean. Imagine then I give you a teaspoon. What you are going to do is fill the teaspoon with water, dump it into the Pacific Ocean, and repeat. It takes 500 billion billion spoonfuls to fill. This is a game that is almost a billion times more complex than Connect 4."

It is, however, a billion billion billion times less complex than chess, a game so complex that even our estimates of its complexity involve guesswork. Unlike chess, checkers seemed to be just—just—within the powers of humans to perfect. "There was a human player, Marion Tinsley, who was world champion," says Professor Schaeffer. "He was idolized. Between 1950 and 1990 he only lost three games. He was as close to perfection as you could imagine a human being. But," he adds, "he wasn't quite perfect. He would make a mistake. It may have been only once every ten to fifteen years, but he would make a mistake."

For many of the checkers players who angrily contacted Professor Schaeffer, they did so out of a sense that his program somehow besmirched the memory of Tinsley. Professor Schaeffer thinks this is silly. "I'm a competitive chess player.

I know many players out there are better than me, but I still go and play because I want to get better. The fact that there is a computer player out there who happens to be much much better than me is irrelevant."

Gradually, though, the checkers community has grown used to the idea of there being an omniscient player—after all, for them Tinsley was practically indistinguishable from perfection anyway. Many have even found a use for the program itself: most notably in solving the "100-Year Problem."

"There was a position published in a magazine in the year 1800," says Schaeffer. "It was white to play, and the question was—is it a win or a draw? A few issues later somebody wrote an article saying it was a win, along with analysis. A couple of years later someone else wrote another article arguing it was a draw." The back-and-forth continued throughout the nineteenth century, until by 1900 the debate had at last ended. "Everyone agreed it was white to play and win."

In 1997, when the program was not even complete, a Grandmaster, Don Lafferty, sent Schaeffer the problem as a nice test of his program. He took less than a second to provide the answer. "It was a draw. Don said, 'How can that be?' I showed him the analysis and he said, 'Oh my God, it's so obvious.' It turned out all the previous human analysis from the third move on had been wrong."

These days, the 100-Year Problem is called the 197-Year Problem.

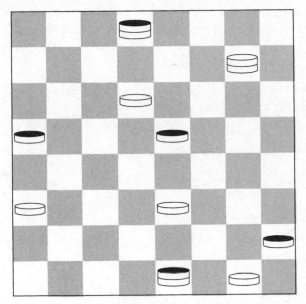

The 100-Year Problem

FACT

Players have not always been obliged to take a piece whenever possible. In the past, if someone refused to jump, his or her opponent could remove the piece as a forfeit and—for obscure reasons—blow on it. Hence this was called a "huff."

APPLE
BOBBING

WHAT'S IT ALL ABOUT?

The least efficient method ever devised for getting one of your five-a-day.

HOW DO YOU PLAY?

Momentarily abandon your dignity, and your knowledge of germ theory, to rummage like a pig in a bucket that has already had several other mouths rummaging in it, hopefully only a small proportion of which had norovirus.

HOW DOES IT END?

With a very cold head, especially if you are playing it at the traditional time of Halloween.

ANALYSIS

Ashrita Furman is currently working on the world skipping record, in the mile category. If he gets it, he will be the holder of "Let me see"—he thinks—"it's over two hundred" records. One of those is the record for the most number of records. Ashrita, in short, has broken a lot of records.

So it takes him some time to remember quite what tactics he employed for one of those two hundred—the most apples bobbed in a minute (he got thirty-four, in 2010).

"The point is training your jaws," he says after a while. "You have to go really wide, chomp down on that apple, and latch." He typically takes at least a month to train for a record and, for apple bobbing—which he enjoys so much he has broken his own previous record—he recommends choosing a summer month. "That's when it's actually enjoyable to duck your head in a bucket of cold water," he says. If you are spending four weeks, working full-time, putting your head in a bucket, such considerations matter.

Under the rules set down by Guinness World Records, he could choose his own apples. If you are able to, he strongly recommends this. "Experiment with the apple. I found harder apples are better—otherwise they disintegrate in the mouth and you come out with just chunks." Equally, though, "If they are too hard you can't get your teeth into them." Granny Smiths are about right; Golden Delicious may be at the soft end.

All of which may well be true. But was it a useful application of his time to reach these insights? "Apple bobbing is silly, I admit, but it requires a lot of focus—you are working at top speed to latch on and get the apple out of the tub. It's a concentration exercise. As soon as you get one, you pull your head out and fling it, and focus on the next apple. I like the challenge of it: it is very simple, just apples and water and you. It's a fun, childlike activity, one of those things like hopping in a sack or running with an egg and spoon you do at a birthday party or fair."

On that note, what about egg and spoon racing or sack racing? Does he also hold records in those? He thinks for a bit. He did, but apparently a lot of serious runners got into egg and spoon racing, and he lost many of his titles. "I think I still have one for fastest hundred meters with it in your mouth." And sack racing? "Yes—I have the record for hopping in a sack fastest over a mile. I did that in Mongolia, racing against a yak. The yak wasn't in a sack." Of course not. That would be absurd.

FACT

Apple bobbing probably began as a fusion of
Celtic traditions about marriage, Christian
traditions about the wanton licentiousness of
apples (witness the Garden of Eden), and—
at the very least—Roman traditions about
actually having apples. Until the invasion of
Britain, the fruit was restricted to mainland
Europe. In the original British game, played
around the world, the apples symbolized
suitors, and unmarried women knelt over the
bucket while trying to latch onto them with
their mouths—an activity with a symbolism
that does not require a sophisticated
understanding of any culture's traditions.

SIX DEGREES
OF KEVIN
BACON

WHAT'S IT ALL ABOUT?

We live in an interconnected world, and Kevin Bacon is the glue. Can you link Bacon to any other actor in the world, just by using films they worked on together?

HOW DO YOU PLAY?

Did you know that Kevin Bacon plays Taxi Racer, the character who beats Steve Martin to a taxi in the 1987 film *Planes, Trains and Automobiles*? Did you know that the year before that, Steve Martin was the sadistic dentist in *Little Shop of Horrors*? Even if you did, you probably wouldn't have noticed Network Exec #1. Which is a shame, because the actor in this seminal role was Robert Arden, who, thirty years before, had been a bellboy (his career sort of plateaued early) in *A King in New York*, a film chiefly noted for being Charlie Chaplin's last acting credit. And, suddenly, you have connected Bacon to Chaplin.

HOW DOES IT END?

Normally with someone trying to connect Bacon to Hitler, or to bin Laden. Which, with the help of Leni Riefenstahl and her documentary filmmaker successors, is actually surprisingly easy to do.

ANALYSIS

Gérard Depardieu is crucial. As is Cary Grant. The key to Six Degrees of Kevin Bacon, says Patrick Reynolds, a computer scientist, is to find "anchor points"—actors who connect between people whose careers don't otherwise overlap. "The hard ones are trying to reach an actor in the 1920s, or in a country where Kevin Bacon never worked," he says.

"John Wayne and Cary Grant get you to early twentieth-century American film." Others, such as Gérard Depardieu, straddle geographic rather than temporal divides. He has a "Bacon number" of 2, meaning that he worked with someone who worked with Kevin Bacon. "In two hops you can be in France." If you have the sort of friends who like northern European films, another good name to know is Stellen Skarsgård. The Swedish actor, who appeared in *Good Will Hunting*, also has a Bacon number of 2—and takes you deep into the bleak realms of Scandinavian cinema. If you really must go there.

For Indian movies, Amitabh Bachchan (*The Great Gatsby* and, er, *Kabhi Khushi Kabhie Gham*) performs a similar role.

Patrick is not actually, he confesses, much of a movie buff. "It's a really hard game. It's a way for people to show off just how much knowledge they have of films." He doesn't have much knowledge of films. His particular talent in this regard is computer science—he runs The Oracle of Bacon, a website that will find the connections, if they exist, between Bacon and any other actor. They almost always do exist.

"Right now the furthest is ten hops. Most people are three hops, but most people you've heard of are two." If, tomorrow, you appeared in a sofa ad then, provided you were alongside someone with a slightly better career than you—who was a jeering peasant, say, in *Robin Hood: Prince of Thieves*—you will probably find you had a Bacon number of 3. Frankly, says Patrick, "even if the advert has a voice-over and he has done other voice-over work, you will likely have a Bacon number of four."

The unnerving conclusion is, much like with rats in New York City, none of us is ever more than a few steps from Kevin Bacon.

Paul Erdős is, in many ways, similar to Kevin Bacon. Both took a lot of drugs in the seventies, both have a prolific back catalog, both worked with some of the finest names in the industry, and both have a parlor game, with the same rules, named after them.

The difference is that while Six Degrees of Kevin Bacon generally involves connecting, say, Stan Laurel with Bacon via the knowledge of an obscure seventies film about a bank heist gone wrong, Six Degrees of Paul Erdős involves connecting Einstein to Erdős via the knowledge of an obscure branch of topology.

Paul Erdős was one of the hardest-working mathematicians in history, with his name on 1,500 papers. Traveling the world from job to job, he would turn up at a colleague's door, take some amphetamines, and work with him or her until they had coauthored a paper, then move on. He was to all intents and purposes a mathematical hobo, but he got away with it because he was brilliant—the arrival of Erdős was a benediction rather than an annoyance, even if he did end up sleeping on your sofa.

Today mathematicians talk about their "Erdős number." An Erdős number of 1 means you authored a paper with him; 2 means you authored a paper with someone who authored a paper with him, and so on. But while a low Erdős number is something to be prized, there is another number even more boastworthy: the Erdős-Bacon number, the sum of both numbers.

Some surprising names have very low Erdős-Bacon numbers. Natalie Portman's is 7: she has a Bacon number of 2, but also an Erdős number of 5, having collaborated on an academic paper when at Harvard. Stephen Hawking's is 7, after an appearance on *Star Trek*. But the lowest is probably MIT math professor Daniel Kleitman, who coauthored a paper with Erdős then found himself hired as an adviser and extra on *Good Will Hunting*, in which Matt Damon played a mathematical prodigy. His number is 3.

In the world of Erdős-Bacon numbers, films about mathematicians generally change everything.

It has taken time but, over the years, Kevin Bacon has come to an accommodation with his eponymous game. When he first learned of it, in the 1990s, "I was horrified," he admitted, speaking at a film festival in 2014. "I started to kind of hear about it in strange ways. People would come up to me and touch me and say, 'I'm one degree!' I didn't really know what was going on.... I thought it was a giant joke at my expense."

Of course, he was sort of correct. The charm of the game is that Bacon has been in a vast number of films but, because he is rarely top billing, knowing which ones requires a certain level of geekery. "I don't think it's a great testament to my ability," he said.

Two decades after it began, the game is still going strong—and Bacon has accepted it is not going to be a passing fad. So, instead, he has embraced it. His charitable foundation is called Six Degrees, in its honor. He also admits he finds the Oracle of Bacon website surprisingly useful.

"Whether it's my age or my misspent youth, sometimes I forget whether I've worked with somebody or not," he said. "I'll look at the call sheet, check the name, and then I'll check their Bacon number. That way I can go on the set and say, 'Good to see you,' or 'Good to see you again.'"

FACT

If you include television documentaries, Osama bin Laden has a Bacon number of 2, making him better connected than most professional actors.

COIN
TOSSING

WHAT'S IT ALL ABOUT?

The arrival of coinage with the Ancient Greeks not only supplanted the barter system, it also made decision making in 50:50 situations considerably more efficient. After all, have you ever tried tossing a pig?

HOW DO YOU PLAY?

Seriously? Oh, all right then. Find a quarter, choose heads or tails (heads is George Washington's head, tails is George Washington's ... er ... arms, actually), throw it in the air, and see which side ends up facing up.

HOW DOES IT END?

With a greater understanding of the vicissitudes of fate. Or, at least, with a greater understanding of who has to bat/break/kick off first.

ANALYSIS

The first thing you need to know, to understand what is almost certainly Matthew Clark's most-cited research paper, is that he is an ear surgeon, and ear surgeons don't much like nose surgeons. They are the Jets and Sharks of surgical specialisms, glaring menacingly across operating theaters, shaking fists at each other in the hospital corridor, spontaneously bursting into song in the emergency room. That sort of thing.

Anyway, one day Matthew realized an opportunity had come up to humiliate his nemeses. "The nose surgeons were doing an experiment," he says. "But their randomizer had broken, so they said they were going to revert to randomizing by flipping a coin." Fools! That's typical ignorant nose surgeon behavior, and Matthew was quick to pick them up on it.

"We said, 'That's rubbish. It is possible to cheat at coin tossing; it's not random.' They said, 'Don't be silly.'" Things, as you can tell, were getting serious: think the Balkans, 1914. It was game on, and this time Matthew would end it for good. He was going to defeat them in the most devastating arena of all: the peer-reviewed journal.

To begin, he needed minions. "We got a load of trainee doctors," he says, "and gave them free pizza to keep them behind after work." Next, he gave them their instructions. "We told them they were to toss a coin, and catch it, and try to get heads every time."

This was not the first experiment in coin tossing. A seminal Stanford paper, "Dynamical Bias in the Coin Toss," had found that all coins are, in fact, biased, not so much by the slight disparities in weight and air resistance that arise as a consequence of different designs on different sides, but by the simple fact that one side starts facing up.

"We prove that vigorously flipped coins are biased to come up the same way they started," the Stanford mathematicians write, in a paper that models the flip of a standard coin, taking into account the fact that it can wobble relative to its spinning axis. They found that their equations implied that just under 51 percent of the time, a coin will show the face it initially started with. Unfortunately, to prove this experimentally, to a statistical level that would satisfy a peer-reviewed journal, would require 40,000 coin tosses, so the researchers were fairly confident that the result would never be verified.

They didn't count on Priscilla Ku and Janet Larwood, undergraduates at the University of California, Berkeley.

Actually, Priscilla and Janet explained, in a write-up of the most dedicated coin-tossing experiment in history, 40,000 tosses isn't that much. "It works out to take about one hour per day for a semester." They got to work—the only concession they made to the scale of the task was, "To avoid tiredness when standing up, the participants sat on the floor." The result? Of the 40,000 tosses, 20,245 landed the flipping side up—a little under 51 percent. The paper was validated.

Matthew's experiment was different, though. He wasn't interested in the intrinsic fairness of a coin toss, so much as whether you can make it even more unfair through manipulation. The only advice he gave the trainee doctors was to practice tossing coins repeatedly, and see when doing so if they could always get the same number of flips.

"It's not that difficult to make it spin the same every time," he says. "Pretty quickly people can learn to make it spin three or four times—enough for a convincing toss—and catch it the same way up." Every single one of the trainees managed to get more heads than tails. "Some people were getting close to 80 percent in their favor. That's after just a few minutes of practice." It seemed the way to reliably bias a coin is just to get the pizzas in and practice flipping it the same amount every time.

The paper was duly published in the *Canadian Medical Association Journal*. The ear surgeons were vindicated; the nose surgeons were vanquished. And that was the end of that.

Except, it wasn't quite over. In fact, several years on, it still isn't. "As a surgeon, you have to do research to progress your career," says Matthew. "You need good research on your

CV." The way research is normally judged is by the number of its citations—the number of people who have referenced the research in their own studies.

The problem for Matthew is, no matter how much quality surgical research he has done, his most-cited paper is on a subject that is tangential at best. "I've been interviewed on radio. I made the front page of *The Telegraph*. I was in six Indian newspapers. It really seemed to catch people's imagination—it has almost certainly beaten my proper papers." Which must be interesting to explain in interviews. Still, at least he defeated the nose surgeons.

FACT

The Tom Stoppard play *Rosencrantz and Guildenstern Are Dead* begins with the pair throwing 76 heads in a row. Assuming they didn't know Matthew's techniques, the chances of this are 7.6×10^{22}—a number about 10 trillion times the population of the world. "A weaker man," says Guildenstern, "might be moved to re-examine his faith, if in nothing else, at least in the law of probability."

DRINKING
GAMES

WHAT'S IT ALL ABOUT?

If you can keep your head when all about you are losing theirs, if you can down a pint from a still-warm sneaker without gagging, if you can remember to point only with your elbow... when the pressure is on—then you have probably been on one too many spring breaks.

HOW DO YOU PLAY?

The games are many, the rules are various. The consequences though are, oddly, universal: an empty wallet, a half-remembered dalliance with a traffic cone, and an extended period of self-loathing.

HOW DOES IT END?

In a police station, a hospital, or a toilet. Or a police-station toilet.

ANALYSIS

"Attempts to develop a rat model of alcoholism have repeatedly foundered," lamented a 1978 research paper, "on the rat's unwillingness to drink alcohol."

There are three problems with getting decent scientific advice on succeeding at drinking games. The first is that ethics committees are oddly reticent about approving alcohol consumption experiments involving human subjects.

The second is that professors of physiology and addiction tend to look rather dimly on those who consider their role as being to give advice on, say, fuzzy duck strategy. "I think this is irresponsible," said one, in response to polite queries about drinking games tactics. "Drinking games kill and should be outlawed." He didn't want his name in this book. Emphatically.

Then there is the third problem: rodent abstemiousness. Most people know the basics of succeeding at drinking games. Line your stomach beforehand; build up tolerance in advance; try, just the once, not to wake up with a kebab on the pillow. But if we want to learn more sophisticated techniques—such as how to maintain enough coordination to rub your belly and tap your head while downing a yard of ale, or how to avoid oversharing in "I have never"—then we are left, as with so much research that has yet to go mainstream, with rats. But what if the rats don't drink?

Thankfully, after that initial difficulty, scientists overcame the problems of rats' natural propensity to teetotalism. "Learned safety increases the rats' willingness to drink ethanol," the paper proudly announced. Over the course of a series of experiments, the rats had been tricked into thinking drinking was safe: they had "learned" safety. This was done by feeding them alcohol, then immediately draining their stomachs, to get them used to the taste without the attendant nausea. Skull and Bones is rumored to take a similar approach.

The path was at last open to scientists who wanted to get rats drunk, and drinking games' tactics would never be the same again.

In the decades that followed, we have learned much about drunk rats. They exhibit what is referred to as contextual tolerance. If you continually give a rat a drink in one set of surroundings, the rat will be better at staying sober in that room, exhibiting fewer signs of intoxication compared to when it is moved to another room.

A subsequent experiment validated this effect in humans. When people drink identical amounts of alcohol, either in a pub or in an office environment, they perform a hand-eye coordination task better in the pub. For those preparing for drinking games, the conclusion is obvious: scope out the location in advance, go drinking in it, and don't compete in your office. The latter condition is probably advisable even without the research.

That is not the only way in which preparation can help. Scientists weren't ready to leave the rats alone just yet—the next batch received not just a drinking room, but a drinking maze. The rats were split into two groups. Half practiced navigating the maze while sober, then had a drink; the other half practiced navigating it while drunk. The drunk rats, unsurprisingly, performed worse.

However, the researchers still weren't finished. They then got all the rats drunk and put them in the maze—and those who habitually navigated the maze drunk suddenly had the edge, even though both sets of rats had had the same experience of traveling in it, and both had previously been given the same amount of alcohol.

The message from this experiment is that whatever the drinking game in question, it is not enough to practice it in advance, you must also do so in context. Before a big night you need to go to a pub on your own, sit in the corner, and down tequila slammers while performing mental and physical tasks. And if that doesn't make you reevaluate your social life and choice of friends, then, well, let's be honest, you're probably a student.

FACT

In Ancient Greek society, the favored drinking game was *kottabos*. This involved men sitting in a circle throwing wine from their glass in such a way that it kept its form during flight and arced toward a target. The exact details of forfeits and rules are hazy, but scholars are united on one point: it almost certainly got rather messy.

MUSICAL
STATUES

WHAT'S IT ALL ABOUT?

It is 4:00 p.m. on the longest Saturday of your life. The cake has been eaten, the presents have been distributed, and there are still ninety minutes of screaming, shouting, running, and breaking valuable ornaments to go until you can hand out the goody bags and offload responsibility for twenty hyperactive six-year-olds to their parents. What do you do? Well, you make them play a game that involves standing motionless for extended periods of time.

HOW DO YOU PLAY?

If you are the child? Stand still when the music stops. If you are the parent? Pretend not to notice the fact that children are very bad at standing still; the longer you can keep them playing, the less time you have to fill before you can sit on the sofa staring aimlessly at a blank wall with a large glass of red.

HOW DOES IT END?

Hopefully, never.

ANALYSIS

After about ninety seconds of not blinking, Chris Clarkson goes through a pain barrier. "The eyes sting, they fill up with water, and a little tear rolls down your cheek." This is a good thing: afterward he finds the urge to blink lessens considerably. "You train the eye. Then it's not exactly comfortable, but it doesn't hurt so much. I still blink, but a lot less."

Most people, even "living statues" like Chris, don't hold

themselves to such high standards. "There are statues that wear sunglasses," he says. Although he doesn't really consider they have the right to call themselves living statues. "To me, that's a massive cop-out." Chris is a purist, when it comes to standing still.

Chris is semiregularly employed to cover his exposed flesh in spray paint, put on clothes stiffened by layer upon layer of masonry paint ("Whoever saw a statue whose toga billowed in the wind?"), and pretend to be carved out of stone.

Generally, whatever the gig, the routine is the same. "You stand still long enough to give the illusion of being a statue. The people around you start talking to each other. 'Is that real, Deirdre?' one will say. 'I'm not sure, Moira.' When they look at each other, and turn their focus away, you move slightly. Even just your eyes. Then they turn back, and go, 'Whoa!'"

Chris is very good at making the world's Deirdres and Moiras go "Whoa!" In 2009, while dressed as a silver James Bond, he won the popular vote at the World Living Statues Festival in Holland. While such an ability might seem esoteric, it is actually a transferable skill, because not only is Chris extremely talented at being a suit of armor or a section of the Elgin Marbles, this also makes him a formidable musical statues player.

For those who wish to emulate him, his chief advice is to know your body's limitations and find ways to circumvent them. "Imagine yourself standing in a bar after a hard day's work. You are constantly leaning on something, putting your weight on one leg or another. It isn't in human nature to stand still; your muscles aren't used to it." So the trick is to find ways to move your muscles without making them do so visibly. "Scrunching your toes up and down is a very good way to get circulation going in your feet."

He prefers a technique that is even harder to spot. "Imagine you have both feet flat on the floor. Put all the weight on the left heel and right ball, then slowly shift it to the right heel and left ball. It's pretty much imperceptible to people watching, but you have completely changed your stance. Little things like that keep the muscles from quivering."

Standing around in a semirigid toga is not completely analogous to musical statues, though—largely because of the "musical" bit. Here, he says, dancing style is key. "Finding yourself in a comfortable position when the music stops is crucial. If both your arms are in the air doing the Y of YMCA, that could be difficult to hold. Think more along the lines of a dad dancing at a wedding. Then whenever the music stops you are pretty much guaranteed to be in a position you can hold."

Where one might need to depart from an imitation of a dad seven pints down at a wedding, though, is in facial expressions. "If you have a massive grin on your face, holding that is remarkably difficult. Facial muscles are not used to holding a position for a long period of time."

But even if Chris is impassive on the outside when working, that doesn't mean he is sad on the inside. "I really enjoy it. Stillness is a very powerful thing to have." At least, normally it is.

"People sneak up on you." He always tries to position himself with his back against a wall; unable to move his head or eyes, he has just his peripheral vision to rely on. "Parents say to their children, 'Go on, tickle him, it's funny,' " he says. "It's not funny."

FACT

The record for the longest time any person has ever spent standing still is held by Suresh Joachim, at just over seventy-one hours. He was allowed to blink.

SNAIL
RACING

WHAT'S IT ALL ABOUT?

Ah, race day. The anticipation, the excitement. The carefully prepared track, the carefully prepared hats. And that one perfect moment when the starting pistol fires, the competitors streak away, and the wind ruffles slimily in the tentacles. Yes, it can mean only one thing—snail-racing season is once again upon us.

HOW DO YOU PLAY?

All you need is a snail, a circle, and a lot of patience.

HOW DOES IT END?

Slowly. Very, very slowly. And slimily.

ANALYSIS

At one point in every serious snail-racing career there comes a watershed moment, a time to demonstrate true dedication. This is when the dilettante snail-racing James Hunts are separated from the coldly calculating snail-racing Niki Laudas.

More specifically, this is the point at which you are required to lick a snail. "I've heard rumors," says Jon Ablett, curator of molluscs at Britain's Natural History Museum, "that the best thing is to lick the soles of them."

A happy snail, you see, is a moist snail. A happy snail is not, though—and this is crucial—a very moist snail. "If you cover them in water they are not too pleased," he says. So the dampness has to be just right. So too does the temperature. "Think when you most often see them out and about—after rain, when it is not too hot and not too cold."

And this is where licking comes in, providing the perfect mix of warmth and moisture. "It gives them the edge," says Jon. The only slight problem is that snails also harbor a parasite—not one that normally finds its way into humans, but then humans don't normally lick snails. "I suppose," he argues, "it all depends how badly you want to win."

The reason snails need to be wet is that they get around by using slime. A study in 2013 followed a group of snails around a garden for three days, mapping their activity. It estimated that a third of all snail food intake gets used up making snail trails. That job is made considerably easier for the snails if it is already wet underfoot. Indeed, in extremis, a dry snail will simply disappear into its shell and wait for it to get wetter.

This was especially relevant to one of Jon's predecessors who, in 1846, received a desert snail shell to add to the museum's collection. Four years later, after a leisurely period convalescing from its transcontinental adventure, the mollusc decided to wake up—the shell contained a live snail after all. It lived happily for another two years.

So even if you are the sort of fair-weather snail racer who balks at bringing the fair weather with you, so to speak, through the application of a warm wet tongue, it is still a good idea to make sure your snail is well watered by other means.

That is not the only way to maximize miles per gallon of slime. The 2013 study also noticed an interesting technique employed by some of the savvier snails. Like Tour de France riders slipstreaming each other, it seemed the more perspicacious gastropods were using existing slime trails by following in the path of other snails. Depending on whether there have been several rounds of snail racing on the course already, it might be worth choosing a starting point at the head of another slime trail, especially if it is relatively straight.

Having released your snail, is that it? Do you then, like a racehorse trainer at the Kentucky Derby, just watch nervously, powerless, and hope they make the running? Absolutely not. Now is the time to motivate, and protect, your snail.

"Most snails have very limited eyesight, so you can't tempt them with colors and shapes," says Jon. The best many can do is sense shadows—which could be threats—and retreat into their shells. So keeping people from blocking their sun is important. Where snail sense really is impressive, though, is in smell. "They have olfactory organs all over their bodies. You want to use smell to pull them quickly toward the edge." In the laboratory he favors mashed-up carrot or cabbage. But also, as many gardeners would attest, "snails do love beer."

Jon came into snails only upon joining the Natural History Museum. "They offered me worms or molluscs." Some might consider that the zoological equivalent of a rock and a hard place, but not he. "I've been here eleven years now and I love snails." He has also come to appreciate their variety—finding niches in almost every land and sea habitat—and has suggestions about how that range could be used to the nefarious racer's advantage. "There's a snail, streptaxis, that is carnivorous. It feeds on other snails," he says, darkly. It looks much like a garden snail. "If you could find one of those you could use it to knobble the competition."

And if it does, if his advice precipitates a great snail-racing scandal, the one upside may be that people could come to understand that snails are more than just dull vegetarians apt to ruin their hardy perennials—and are worth saving. "There have been more land snail extinctions in the last hundred years than in any other group," he says. "When something doesn't have whiskers or pretty feathers it is hard to make a public case for funding conservation, but they are a really important part of the ecosystem," he says. "They are fascinating creatures, sadly ignored."

Nevertheless, even his love of snails only stretches so far. He has raced them just once in his life and, an embarrassing admission for a snail expert, he lost. Did he lick his snail, though? "No I did not." Well, there you go then.

FACT

The world snail-racing championships, held in Congham, Norfolk, since the 1960s, began as an import from France, after a resident witnessed a race there. Although farmer and "snail master" Neil Riseborough has said that it might have been a cultural misinterpretation of what was going on. "It may just have been some snails that were running away from a cooking pot."

COMPETITIVE
EATING

WHAT'S IT ALL ABOUT?

Your reaction to all-you-can-eat buffets is crucial. When invited to go to one, do you:

a) Think, "This boyfriend isn't a keeper...."

b) Carefully consider whether or not your current level of hunger justifies the expense. No one wants to be one of the fools who subsidizes it for everyone else.

c) Prepare yourself for battle.

HOW DO YOU PLAY?

In the case of an all-you-can-eat buffet? Well, you try to exceed five and a half pounds of assorted food in twelve minutes—the current buffet record, held by solid all-rounder Crazy Legs Conti.

HOW DOES IT END?

Probably with indigestion. Possibly with worse. Certainly with a total loss of dignity.

ANALYSIS

Dieters have long relied on a particular physiological quirk to control their appetite. "The brain does not receive word from your stomach that it is filling up until ten or twenty minutes after you start," says Crazy Legs Conti. This means that by eating slowly, you are less likely to overeat. Crazy Legs is not a dieter. People called "Crazy Legs" rarely are. But he still finds this advice helpful.

"It tells me that you have got ten minutes of really frenzied eating before the brain knows what's going on," he says. "That's how you get one over on the lucid, rational part of the brain." Crazy Legs spends a lot of time thinking how to get one over on his brain or, in his words, push it "past the point of society, of sanity." Because, since 2002, Crazy Legs has been a competitive eater.

It began for him in a bar in New Orleans while his friends were watching the Super Bowl. He was trying to console himself after having not gotten tickets. It was an oyster place. It was also, perhaps, less classy than that description sounds. "They had a record for oyster eating—33 dozen. If you beat that you got them all for free."

Now, this seemed like a challenge. Crazy Legs had always enjoyed seafood. Indeed, "it's as if my stomach is a beacon welcoming the ocean's creatures." So he sat down to see if his stomach could welcome in 400 of them in one sitting. It was difficult, but he did it. And word spread; a new talent had entered competitive eating.

Even now, after more than a decade, several world records, and consistent success in the extremely specialist corn-on-the-cob discipline ("It's strange to think of my parents saying, 'Eat up your vegetables,' and having it lead to being four-time corn-on-the-cob champion of the world"), he is not a big man. Winning, he says, is less about body mass than about a positive mental attitude. "It's almost always unpleasant at some point," he says of his job. "Ultimately, success is about mind over stomach matter."

Some of his colleagues ("My brothers in stomach") use sports psychologists. One "attaches marshmallows to a string and hangs it in his mouth to try to control the gag reflex." Another, Joey Chestnut—the biggest name in hot dogs—

likes to use hot water. "He thinks it loosens the internal sphincters. All but one, hopefully."

Crazy Legs just believes there are no shortcuts. As in any sport, it is about hard work and discipline. Each food requires a particular technique (see box on page 146), and he tries to learn it. "Are great eaters born or made? I only got better at basketball by playing on the court by myself with a ball and a basket—by instilling muscle memory so I don't need to think. The same is true with competitive eating. You have primed this human machine to become a woodchipper, to break down food and perform Tetris in your stomach."

Eventually, it should become automatic. "If you don't need to think about the next bite, about the next hot dog—that's the mental state you need to be in. With corn on the cob you could be halfway through your meal, 25 cobs down, and think you are doing great—but the moment you pause to look at it, that's when it's over."

Between competitions he eats sensibly and exercises. "I work out: I jog, I run marathons. I used to do yoga. I'd imagine true yogis would probably feel pretty repulsed that I do it so I can eat more hot dogs." Then in the run-up to a big contest he modifies his diet. "I have a lot of Japanese soup— it is easy to digest and fill up on." Finally, on the day itself he doesn't eat at all and instead works on getting himself in the right mental state. "Your stomach needs to get ready. The big intestine needs to drop anchor to the lower intestine." Then he is ready to go.

He knows that the hours that follow will be crucial. Recently, he lost a hot dog eating competition by a quarter of a bun. It was in his mouth, but he couldn't swallow. "What is in the mouth at the end counts but you have to finish in a timely manner." He couldn't.

This does seem a little extreme. "In terms of physical feel, you almost always feel like you are bursting," he admits, "but you have to hold it down." Major League Eating, being a civilized family activity, frowns on onstage vomiting. "By the time you are finished, have done some interviews, and spoken to fans it is a few hours later and it has had time to settle. It's the next day, when you wake up, that you really realize what it is you've ingested. Sometimes I'll wake up and think, oh, that's right, that's what 100 chicken wings feel like."

This is when, perhaps, he has a brief moment of introspection. "It's a little strange to think my mind could be doing other things. It could be working on a cure for HIV, or creating an opera. But no, it's going to focus on eating 100 chicken wings." A pause. "But the world has got plenty of operas."

THE DISCIPLINES

Hot dogs
Target: 69 in
10 minutes

"Most people separate the bun and dog, and dunk the bun in water. Some people do two dogs in one, and some eat them together with the bun. That's a bit strange—it's called 'picnic style.' Basically, it's about getting into a rhythm, ignoring the guy beside you, and making every chew and swallow work. You want to have dogs left at the end—because you can stuff them in easily. Buns are a problem."

Corn on the cob
Target: 46 in
12 minutes

"I use the manual typewriter approach. Left, right, left, swallow, repeat. I try to get it in five rows or, if it has been a bountiful harvest, six. Swallowing is important—I don't make a dinging noise like a typewriter would but I do flap my arms. You have to remind yourself to swallow. Largely, it's about jaw strength and focus."

Chicken wings
Target: 182 in
30 minutes

One of the more technical disciplines, this has several successful attack strategies and there are many videos online detailing different techniques.

i) Meat umbrella. "I push it down, it forms an upside-down umbrella, and I pop it in."

ii) Wishboning. Put whole wing in mouth, while holding the end of the small and large bone. Break them apart and pull, sucking the flesh as the bone leaves.

iii) Cluster targeting. "Like a cannibalistic chicken, I look for the meaty parts, and I just peck at them."

FACT

Perhaps the first eating competition recorded in history was between Loki, the Norse god, and Logi, a giant. Loki made a strong attempt, devouring a plate full of meat, but Logi outclassed him—by eating the plate as well.

QUIZZES/
TRIVIAL
PURSUIT

WHAT'S IT ALL ABOUT?

I say "Norma Jean Mortenson," you say "Marilyn Monroe." I say "ice pick in Mexico City," you say "Trotsky." 2.71828? e. Lima? Peru. And, the clincher: what if I say "pub"? If your answer is "arena of gladiatorial combat for proving one's social supremacy through the recall of unusual facts," congratulations—you're a quizzer. If you answer instead, "pleasant social place to meet friends, drink, and gossip," please, please just don't talk during the music round.

HOW DO YOU PLAY?

Choose a team name that starts off as mildly amusing and becomes progressively less so as the evening progresses. "Let's Get Quizzical" works very well, as does—for those quizzes in which the quizmaster reads the score out each round—"I lost my virginity at the age of…"

HOW DOES IT END?

Hopefully, after the final round, with the quizmaster at least admitting to having been past the age of consent.

ANALYSIS

"There is this strange idea," says Olav Bjortomt, "that quiz knowledge is innate." It isn't. "It takes hard work. Dr. Johnson said a friendship is in a constant state of repair; the same is true for general knowledge."

For Olav, member of the England quiz team and 2014 European Quizzing champion, this constant repair requires constant vigilance. "I scan adverts on the London Underground to see what's out." Most recently, that included an ad for the film of the book *Fifty Shades of Grey*. "So I thought, 'What is the real name of E. L. James? Who played Christian Grey?' There are always lots of related facts."

As well as ads, every day he looks through newspapers for possible questions. "Quizzers know what to learn. I am very good at recognizing proper nouns when I read the newspaper." Sometimes, it just requires spotting the likely questions. "You hear facts instantly and know they are quiz questions. Like, Harper Lee has a sequel out to *To Kill a Mockingbird*, what is it? That's a quiz fact." Years ago, he set a quiz and put in, "'What is Lady Gaga's real name?' Now I hear that in quizzes around the country." That is another quiz fact.

Other times, though, succeeding at quizzing involves active learning. He knows all the Shakespeare plays, obviously, but he also has to have a passing knowledge of the other Elizabethan playwrights. "With Ford, you have to know *'Tis Pity She's a Whore*. With the others you learn the most famous play and one other." He takes a similar approach elsewhere. "I learn a couple of moons of Jupiter, a couple of chancellors of Germany. I can't do more, or I get bored."

This methodology cuts to the central, inescapable truth about quizzing: it's about knowledge.

"It is like buying tickets for a lottery. The more lottery tickets you have, the more chance there is of winning." He has friends who have gone so far as memorizing every Trivial Pursuit question (see box on page 153). "It might seem like that is trying too hard, and a little sad, but that's the way you do it."

It is this dedication to increasing his number of lottery

tickets—extending even to German chancellors—that has enabled Olav to make quizzing not just his hobby but his career. In his day job he is a quiz setter for television and *The Times* of London. But the talent became lucrative long before he found formal employment.

"I did pub quizzes when I was sixteen or seventeen. We bought one drink all night and had people chasing us out of the pub after we won three or four nights in a row." At university he was cleverer, and went to a selection of pubs. "We made $235 a week. You can't do it on your own, though—I would feel slightly resentful of my friends making money off me. But that's what friends are for, I suppose."

Indeed. And perhaps they earned it in other ways. Because, to further paraphrase Johnson, his friendships are in a constant state of disrepair. "I realize I do mentally delete a lot of everyday information—stuff about friends and their lives," he says, "in favor of quiz facts."

There is only one way to guarantee victory at Trivial Pursuit: learn every single answer. And actually, says Ed Cooke, a memory grandmaster, that's not as ridiculous a proposition as it sounds.

A typical Trivial Pursuit game has 400 cards, each with questions in six categories. "Each of these is a paired fact," says Cooke, who now runs a website, Memrise, that applies memory techniques to education. The first fact is in the question, the second is in the answer. "To learn them, you first need to visually link the question to the answer."

He chooses a question at random as an example. And, given the visualization that will follow, it is worth emphasizing this really was chosen at random. "Q: What punk group did Handsome Dick Manitoba sing for?" The answer? As I'm sure you all know, "The Dictators."

"The thing is to form a good image to link one side to the other. Dictator is easy—imagine a dictator onstage." Hitler, for instance. Next, you have to have Hitler doing something that relates to the question. "You have to link Hitler to … Handsome Dick."

Once you have done this, and I think we can all agree there are several options, it is time for the hard work. "Then you are in the business of testing." This is not just about repeating "Dictators, Handsome Dick Manitoba." It is about being asked the question and actively recalling both the image and the answer. "This hugely strengthens the technique."

Finally, you repeat the process. "Go through it again in an hour, then that night, then next week. The spaces get bigger with time and it is really efficient—you are reviewing your knowledge just before it is about to drop from memory."

How long should all this take? There are 2,400 questions. Assuming you have a reasonable level of general knowledge and know a third of them already, if you take an hour a day on the rest, split into chunks, Ed thinks you would be ready in a month.

Consider it an Advent penance, before a well-earned Christmas Day triumph.

FACT

Think you have what it takes to be a professional quizzer? Below are some questions from the European Quizzing Championships. Answers are in Appendix 1 at the back of this book.

1. Which soil insecticide was widely used from the 1950s to the 1970s to control root worms, beetles, and termites until it was banned in 1974? It shares its name with the man nominated for Best Cameo at the 2011 Scream Awards for his role playing himself in *Transformers: Dark of the Moon*.

2. Named by Christopher Columbus after his brother, which French island was a Swedish colony for a significant length of time, still reflected in the name of its capital, Gustavia?

3. Which mortal angered the goddess Athena by beating her in a weaving contest?

DIPLOMACY

WHAT'S IT ALL ABOUT?

Do you find that Monopoly is all too brief? Does a game of Scrabble with extended family do too little to foment deep and lasting antipathy, even when you keep overruling other players' words with references to the *Oxford English Dictionary*? Well, how about a game based on pre–World War I Europe that is even more likely to produce bitter and endless stalemate than that conflict, and that actively encourages shafting your friends and family? Excellent. Diplomacy is for you.

HOW DO YOU PLAY?

Not that diplomatically, to be honest. A little like Risk, but with none of the elements of chance, it is about the total military domination of Europe. There are no dice here; all you have is cunning and guile. Each move takes 15 minutes and requires negotiating with, or lying to, the other players. If you want to take the Dardanelles without suffering your own Gallipoli, just like in real life you are going to have to woo—or deceive—the Ottoman Empire.

HOW DOES IT END?

It'll all be over by Christmas. Probably.

ANALYSIS

"Sometimes, in diplomacy," says Simon Rofe, lecturer in diplomacy at the School of Oriental and African Studies, University of London, "you have to talk to really unpleasant people."

The British government sat down with the IRA. Israel signed the Camp David accords with the PLO. The Afghan government negotiated with the Taliban. And, come 1:00 a.m., amid stalemate on the Franco-German border and despite the fact that the dirty, underhanded cheater pulled a fast one on you in the Gulf of Bothnia, it could be time to call a summit with your sniveling uncle Charles.

Simon likes to use Diplomacy (the game) to teach diplomacy (the subject). Its chief virtue, he says, is that it rarely ends. "In some senses that is the quintessential thing about it," he says. While many might question the purpose of a game that almost never yields a victor, Simon says it demonstrates an important point. "Cardinal Richelieu established the French foreign ministry, and he did so on the basis of *négotiation continuelle*. You can't just have a war, have a peace conference, then not speak to people for the next hundred years. You need someone in situ, to represent the head of state and be able to further your interests continually."

Even for the stronger nations, Diplomacy teaches the value of keeping on good terms with everyone. Otherwise, "You can win the battle but lose the war.

"Someone can clearly be in a position of power. That doesn't mean Belgium does not have influence. The game can bite you if you overlook the small players. If you just concentrate in one direction you can easily be ganged up on, lose your base, and find yourself traipsing around Europe like Hannibal.

"I've seen the game played skilfully by people who just want to maintain a position on the board. They move around Benelux.

They might lose Luxembourg, gain southern Belgium—for them, longevity is the quality, and to overlook them means someone else might bring them on their side. You don't want to be isolated, even if you are the biggest player. Otherwise you end up in the US position, being a unipolar power without friends."

To avoid this scenario means understanding the art of negotiation, even with people you distrust.

"You don't have to like each other," says Simon. "What you need to know is where his interests lie, and where yours are. Previous form makes a difference. Have you got a track record of being victorious, or trustworthy? Do you do things when you say you will? Have you said you will attack and done it? Have you said you will negotiate all the way but done an underhanded deal instead?"

If so, if you are in the position where trust is irrevocably lost between you, that is when you need an intermediary—a Jimmy Carter figure. "Have you been able to engender friends elsewhere? I may not trust you, might never trust you, but if we share a friend we can still do a one-off deal—a bilateral trade."

And once you've convinced Uncle Charles to do that, that's when you can do a secret deal with your credulous cousin Matthew and wrest back control of the Gulf of Bothnia.

In high-level diplomatic negotiations, one should never underestimate the importance of baser human needs. In particular, of sandwiches.

In the mid-1970s, Sir Christopher Meyer, former British ambassador to Washington, longtime Diplomacy fan, was part of a delegation to Moscow. His team's job was to thrash out a communiqué over the weekend that the respective leaders could release on Monday.

"We were invited by the Russians to begin negotiations after a nice lunch at the hotel," he says. To him, it sounded like an excellent suggestion. More experienced hands suspected something was afoot. "My friend said, 'They're up to no good.'" Just in case, he got the embassy kitchen to make them some sandwiches, which they hid in their diplomatic bag.

"We arrived and they said, 'Before we have lunch, let's do a bit of work.' My friend looked at me and said, 'Here we go.'" Of course, lunch never came.

"They thought we'd crack by tea."

They had not reckoned on the diplomatic bag. "At three p.m. Julian said to me, 'Get the sandwiches out.' We started chomping merrily on ham sandwiches. The guy on the other side was a survivor of the siege of Stalingrad. He smiled, he had a sense of humor."

"The negotiations went into the next day. But they were saved by sandwiches."

So while the future of Europe, or at least the fantasy Europe in your game of Diplomacy, may seem to depend on the complex web of alliances between the central powers, in reality it may have more to do with whether your opponent is hungry—or, equally likely, drunk. It was an important lesson for Sir Christopher to take into his other diplomatic work—the work that took place in a dacha just outside Moscow.

"The embassy had a country house." As well as being part of visiting delegations to Moscow, he was also stationed there for a time. "During the dark winters I would rent it for a weekend, go down there with eight or nine friends, and play Diplomacy." The games would go on until late on Friday night, then continue for most of Saturday until "by three a.m., usually, the game would have collapsed." If it didn't, they resumed on Sunday.

As one of a vanishingly small group of people to have successfully completed a game, Sir Christopher believes he has spotted useful patterns.

"Certain countries always won: they tended to be big. Germany, the Austro-Hungarian Empire, Russia. The fringe countries were Turkey and Britain. Turkey used to mimic real history—it would penetrate the Balkans, get into Austria, and be pushed out again."

Nevertheless, Sir Christopher, a patriot, tended to choose to play as poor weakened Blighty. "I always had great hopes for a British–Turkish alliance, with a pincer movement on the other countries. There was usually a betrayal by one, or both, though: they would go off with someone else."

So it was that, despite one of the stellar diplomatic careers of the twentieth century, he admits he did not have similar success in the board game version. "I loved it, I loved the game. But the corollary between playing it and doing it is not that great," he says. "In the real world, alas, I never got the ability to offer people bits of territory or declare war."

FACT

Of all the great diplomatic gaffes—and there are many, from JFK calling himself a cream bun in Berlin to George H. W. Bush vomiting on the Japanese prime minister—one of the favorites among the community concerns the 1960s Labour Foreign Secretary George Brown requesting a dance at a drinks reception. "I shall not dance with you for three reasons," reportedly came the reply to Brown's chivalrous offer. "First, because you are drunk; second, because this is not a waltz but the Peruvian national anthem; and third, because I am not a beautiful lady in red, I am the Cardinal Bishop of Lima."

SANDCASTLES

WHAT'S IT ALL ABOUT?

In medieval times, the castle was a home and a refuge—but also a symbol of power. The most elaborate of all rose from the desert sands of Arabia: Crusader forts that kept whole nations in check. These days, rising from the sands, a new breed of castle—just as elaborate, if a little less robust against catapults—keeps beachside competition in check. Can you construct the finest sandcastle and see off other would-be seaside fiefdoms?

HOW DO YOU PLAY?

Look at a picture of Krak des Chevaliers, or Carnaervon. Fill your bucket and get to work. End up with a small mound. Exclaim loudly that you are pleasantly surprised with the way your motte and bailey has turned out.

HOW DOES IT END?

With high tide.

ANALYSIS

Most people get it wrong from the beginning, because the first mistake, when building a sandcastle, is to use a bucket. "The sand just doesn't come out. It's airlocked," says Matt Long, a professional sand sculptor.

Anyone who has carefully packed a bucket then turned it upside down will know the feeling. The sand forms a vacuum with the base, and stays stuck on by suction. So you shake and you prod and you tease, and you bash an upside-down bucket with a spade, and even before your sandcastle has begun you have undermined its structural integrity.

Instead of the sort of imposing late-medieval crenellations and formidable drawbridge that will see off the puny castles of your beach competitors, you end up with something that would shame a moderately ambitious Iron Age chieftain.

The solution is simple, says Matt. Take a bucket and "cut the bottom off." When he builds sandcastles, he makes constructions so big that they would not do that badly as real castles. He packs the sand in tight and uses lots of water and, crucially, he does so in situ. In that way, rather than having to turn your bucket upside down and then coax out the sand, you just pull the sides up and reveal a perfectly bonded shape, already attached to the ground below.

Matt, admittedly, uses large wooden sides to make his molds rather than red plastic buckets with the bottom chopped off, but for those not building 12,125-pound, 17-foot-high fortresses (a recent construction) he has an alternative suggestion.

"Take a big paint pot, remove the bottom, stand it on ground, pack it with sand and water, and slide up the sides—then you have a block of sand to cut," says Matt. Get some successively smaller paint pots and you could even build a tower, one on top of the other. "That's the very basic way to create elevation in sand. Because that's where the drama is—to make sand do what's not expected of sand. When sand is standing up two, three, ten feet, that's when it gets dramatic."

And what if your sand does not stand up that high, whatever you do? Well, it might not be your fault: different sands can have varying qualities. "With some sands you can carve out seven feet of sheer wall. With others, you are lucky if you get ten inches." The best sands are quartz crystal, because the individual grains lock together. The worst sands are desert sands—worn and ground down into spheres by millennia of erosion. In the Middle East you can't even use the beach sand in the construction industry. When Matt sculpts there, they have to ship in his sand from abroad.

Once you have a block of sand standing on its own, all that remains is to take away the bits of sand that aren't castle. Matt's tip, one surprisingly frequently ignored, is to "always, always" start at the top and work your way down. Otherwise your finely crafted arrow slits will find themselves crumbling from the sand tumbling from above.

Oh, and finally, lest you think it is time to branch out from medieval fortifications, Matt's advice is: don't. Michelangelo didn't say, after *David*, "That's enough of perfectly honed marble bottoms and asymmetric testicles, I'm going to try something more avant-garde." The Polynesians of Easter Island never thought, "Maybe our small island is full up, as regards stylized stone heads?" No, they knew their market. So it is with sand.

"Sometimes," says Matt, "I carve something artistic. Always, I find people say, 'What is it?' I reply, 'It's art, dammit.'" Then, resigned, he goes back to castles. "Everyone understands a castle."

The sandcastle created by Maryam Pakpour is not, at least conventionally, impressive. There are no battlements or arrow slits, nothing in the way of decoration—and not even a moat. Yet, for her paper in *Nature's Scientific Reports*, the Iranian scientist has constructed something so outlandish that at first you assume it must be Photoshopped.

Her sandcastle is 60 centimeters (almost two feet) high—impressive, but not notable. What marks it out, though, is it is only seven centimeters (not quite three inches) wide. Among sandcastle enthusiasts it is—and one doesn't use such a phrase lightly in the dynamic world of beach sculpting—a paradigm shift. An Ancient Egyptian, comfortable with the idea that buildings can be high, but only if they are also wide, would doubtless experience the same shock on seeing a modern skyscraper.

Maryam's tower stands unsupported, impossibly erect. Less a castle it is more a lighthouse, beaming a clear message to all beachside sandcastle competition: this is a new kind of sand sculpture. It also beams a clear message to physicists—whether or not they are beachside competition. Because such a sandcastle should not be possible.

Conventional physics had it that a straight-sided sandcastle could not exceed much more than 20 centimeters (seven inches) in height. Having visited a beach, Maryam concluded this estimate was "in stark disagreement" with what physicists like to call "reality." What was going on?

She and her colleagues re-estimated the theoretical limits of sand towers, accounting for their ability to bond using "capillary bridges." For the same reason that water forms a meniscus, climbing up the sides of a container using surface tension, it also creeps between grains and pulls them together. "This then creates a network of grains connected by pendular bridges and allows, for example, creating complex structures such as sandcastles."

Maryam went back through the equations and came to a conclusion that better matched what actually happens on beaches. Instead of a theoretical upper limit, she worked out the relationship between minimum base diameter and maximum height. In theory, according to her calculations, a cylinder of sand 20 centimeters in diameter could reach a height of 2.5 meters (8.2 feet). But what about practice?

In one of the more unusual "methods" sections of a *Nature* paper, Maryam described how to turn the equations into reality.

"Cylindrical 'sandcastles' were constructed using non-wetting PVC pipes of different diameters cut in half over the length of the tube. The two halves were assembled, and the wet sand was put in the tube standing vertically on a surface." It was then compacted by bashing vigorously on the top. "This process was repeated until the pipe was filled with sand up to a certain height. The two halves of the cylindrical tube were then carefully removed."

Maryam beheld her creation and, her paper recounts, realized she "had the recipe for the perfect sandcastle."

It was a single symmetrical tower, pointing to the sun while, around its base, boundless and bare, the lone and level sands stretched away. Look on her works, ye sandcastle builders, and despair.

FACT

Sandcastles are at least as old as the Greeks, and probably as old as castles. Heraclitus said, "History is a child building a sandcastle by the sea, and that child is the whole majesty of man's power in the world."

STONE
SKIPPING

WHAT'S IT ALL ABOUT?

A life led by desire, said Kierkegaard, is like "a stone skipping across the waters that skips the surface of the abyss only to sink to the depths the moment it pauses in its flight." What, though, if what you desire in life is to skip stones? What if you get so good at it they can bounce dozens of times and amaze onlookers? That suddenly sounds like rather a good life. Even Kierkegaard might be cheered by that.

HOW DO YOU PLAY?

Carefully select a rock. Throw the rock. Watch it bounce once then crash into the waves. Blame the rock.

HOW DOES IT END?

No matter how high the spin or how fast the throw, stone skipping, like political careers, must end in failure—eventually everything disappears noiselessly beneath the water. Maybe Kierkegaard had a point after all.

ANALYSIS

One day in 2000, Kurt Steiner came across a small article in a newspaper advertising a stone-skipping contest. He showed it to his wife. "I said, 'I used to be pretty good at that.'"

Kurt had grown up near the shores of Lake Erie, where, for hundreds of miles of coastline, eons of geological time has ground the stones into perfectly rounded disks, perfect for skipping. "Anyone who lives there just does it naturally."

"My wife said, 'Then let's do it.'" Kurt had a condition, though. First, they would have to get Erie rocks, from 100 miles away. "I'm a purist with rocks. I had to get those rocks, and haul them all the way to the competition." She agreed, and Kurt and his Erie rocks won the amateur event.

There are two responses to this story. The first is to note that if there was an amateur event, that implies there was also a professional one—for stone skipping. There was. The second is to query how Kurt found such an amenable wife.

On the first point, Kurt, who now does consider himself professional, concedes it is "not a lucrative business." On the second, well, just you wait.

Over the following years, Kurt began to make a name as one of the best stone skippers in the world. Between 2002 and 2006 he held the record for the most bounces. Then in 2007 his world came crashing down. Someone wrested the title from him with a stone that skipped more than 50 times.

"That triggered it. I realized I had to unpack everything I knew and start like I did not know anything. I had to try every nuance and concept." That was when he explained to his wife that he was going to quit work to become the finest stone skipper the world had ever seen.

The five years that followed are best understood by viewing them as an eighties movie montage, from *Rocky*, say—but with a wife occasionally in the background wondering why "Eye of the Tiger" is playing and her husband is reading books about stones.

Kurt began training his body and honing his mind. He studied the scientific literature and analyzed his opponents' techniques. "I went deep into physics analysis—into geometry and force considerations. I started doing things with the stone that I have never seen explained in the literature."

There is a significant body of work on stone skipping. Most physics papers advise the stone is best angled to hit the water at a shallow angle. They offer advice on its tilt, its internal inclination, and its spin. But, says Kurt, they are all flawed.

"The models assume an idealized disc, in an idealized situation, with no air resistance. But the stone doesn't behave in that way when the disc is not symmetrical, the water surface is not flat, and there is air resistance." All of this means in the real world, stones do not behave in what he calls the "classical" style.

"In the classical style, you want the first bounce high because a classically skipping stone is very much like a ball bouncing on a surface. Every bounce is lower than the last, so if the first is high the others will be, too." This analysis would have you tilt the stone at an angle so that it bounces up after the first hit.

The first clue that this might be incorrect came from listening. "You can tell when you throw a good combination of spin and speed because you can hear it—the stone sounds like a bird fluttering." Air resistance is not negligible at all.

"When I throw, I want the first bounce to come out almost horizontal. I aim to go in at thirty degrees and come out at five. Instead of, say, going in at fifteen and coming out at ten. In this way more energy is captured in the forward movement of the stone. As soon as you add vertical oscillation you also lose a tremendous amount of energy to air resistance."

Air resistance can also have its uses. In conventional models, the stone is assumed to be thrown in a parabola, pulled down by gravity and curving to its first contact with the water. He has found that his stones sometimes curve almost the other way. Such is the speed and spin that, "in practice, wing and propeller forces can lengthen the travel path as the stone glides into contact. It depends on the shape and weight of the stone, but even an aggressively steep throw can impact at angles closer to the horizontal." The stone is quite literally flying.

"By embracing and harnessing complexities, I've been able to make a stone skip in ways that existing descriptions do not even consider. At least, that is my feeling; I have not had my specific concepts reviewed or tested by a professional. I've asked, of course, but it's like they have better things to do!"

There are five variables you can control in stone skipping, of which the angle of entry is just one. The others are simpler, though, at least in theory. There is the stone's own tilt—how level it is—which in Kurt's model should be kept close to flat, but never pointing down. There is its "twist"—its left–right inclination. This also needs to be flat. Then there is spin and velocity, which both need to be as high as you can make them.

It is achieving the latter that is the hard part. "Making it impart the particular forces you need is not simple." This should not be surprising. "If you get someone off the street to throw the hammer or discus, they're not going to be able to do it. It's about training your body to a very specific action." He says most of his colleagues have had sports injuries, including some necessitating back surgery.

After his five years of investigation, Kurt realized there were two styles that work: whip and drive. He is a rarity in that he can do both. "Drive is more like a baseball throw. A lot of strength comes from the back legs, lunging forward and low. In whip you are contorting more, throwing the stone down at the ground almost at your feet." It is with whip that he uses a thirty-degree angle rather than the shallower approach that the literature recommends. "The shoulders are rotating in such a way that they make a whip out of the arm—at the last second your arm forces that stone forward." He demonstrates both throws in a YouTube video titled "Whip vs Drive."

Both approaches have their merits. "Whip amplifies spin. Drive amplifies velocity. With drive you can throw a heavier rock faster, but with not so much spin." His main piece of advice to novice skippers is to just experiment, to find a style that works for them, as he did. "Don't be afraid to do something that feels uncomfortable. If you only do what feels natural you may not hit on the best actual method for you."

He had learned to throw by using drive, but on the day Kurt felt ready to try for the world record, he chose whip. It is worth watching what happened next on YouTube.

This throw was Kurt's gift to the world, his Sistine Chapel. The rock seems less to bounce than to float. Lifting off the water after the first impact, it temporarily takes flight, then returns to break the surface of the lake in batches—rat-a-tat-tat, rat-a-tat-tat.

On that day, everything went right—forty bounces, fifty, and still it continued. It was for throws like this that Kurt got into stone skipping. "The great thing is it's so primal," he says. "In this world, where everything is so connected and everybody is so focused on some device in their hand, I find it essential to replace that with a piece of the earth, a rock.

"There's nothing more basic or satisfying than that. Basically, you're making a stone float. You're in control of that magic. It's like watching a fire or a sunset; it's putting a person in physical contact with something millions of years old." In all those millions of years, though, no one had done this.

Sixty bounces, then seventy. A slight corrugation of ripples on the lake helps it far past the previous record—the undulations seemingly coming up to meet each bounce. It doesn't so much rebound from a flat surface as slice off the tops of successive peaks. Eighty bounces, then rat-a-tat-tat, eighty-eight. The stone reaches skipping apotheosis, seemingly giving up on bouncing at all and just sliding along the water instead until, victorious, it at last sinks beneath the surface.

It was as close to an unbreakable record as most stone skippers could imagine. All Kurt's work had been worthwhile—for him, at least.

Did his wife agree? Well, he says, these days he is in semiretirement, and when he talks about skipping, "she says, 'Hey, I thought we were done with that now?'" She is, he says, "inscrutable ... I think she likes to see me successful in the end. On the way there she has been a little bit long-suffering. She is ..." he thinks of a better word, "tolerant."

FACT

Stone skipping is not just a harmless pastime; it can also be used to sink ships. A sixteenth-century book, *The Art of Shooting in Great Ordnance*, by William Bourne, has a chapter titled, "How and by what order the shot (doth) graze or glance upon the land or water." By the eighteenth century, such grazes were an accepted tactic in naval engagements. One manual, with a liberal interpretation of the word *playful*, refers to how "at sea, as on land, cannon balls effect greatest damage not when striking directly the object aimed at but when rebounding in 'playful' mood from land or water and 'bowling along.' This is called ricochet practice."

TWENTY
QUESTIONS

WHAT'S IT ALL ABOUT?

Animal, vegetable, mineral, or game? Game.

Does it involve a board? No.

Does it lead to intense and persistent feuds? Not normally.

Can you do it outside? Yes.

HOW DO YOU PLAY?

Does it require a stick? No.

Is it stone skipping? No.

HOW DOES IT END?

I give up. It's twenty questions.

But you don't do that outside. You don't, but
you can if you want.

You're an idiot.

ANALYSIS

Robin Burgener has a pretty good idea about what you are
thinking of when you play twenty questions. "You can almost
guarantee it," he says. "People new to the game will go for a cat,
or a dog. Then they will think of a vegetable—usually a carrot."
Amazed that Robin's online computer program manages to
guess both, they will then resolve to stop going easy on it.
"They'll make up something really difficult, like a duck-billed
platypus. Then they're blown away when it gets it in ten goes.
But, really," he says, "how many egg-laying mammals are there,
covered in fur?"

In 1988 Robin designed a computer program to play twenty questions. People using it think of an animal, vegetable, or mineral and, just as in the parlor game, it asks questions in an attempt to guess it. These days, it normally succeeds.

When it started, though, it was only in extremely specific circumstances that it would defeat the human. Namely, if the human was thinking of a cat. "All it knew was that," says Robin. Slowly, over time, it learned from its losses—starting with the most common subject. "Next came a dog, then a carrot." Then, presumably, a duck-billed platypus.

These days, twenty-five years later and after being made into a website and an app and used for millions of runs, how many things does it know? Well, a bit over 5,000. Out of all of the world's billions of animals, vegetables, and minerals, it seems, people who play twenty questions are only ever choosing one of 5,000 of them. Even when people are trying to be obscure, they are obscure in a predictable way.

In fact, it's even more limited than that. "Ninety percent of games draw from fewer than 100 objects. The top 1,000 objects gets you one or two percent more."

In theory, this means 90 percent of games should end in seven guesses: the most efficient search strategies, binary searches, eliminate 50 percent of options each time. If you start with 100 options, after one guess you have 50, then 25, and so on. So, to play twenty questions successfully, when looking at all the likely objects remaining, each guess should try to get rid of half of them. This is not the way a human works. "When guessing, a person narrows it down quickly to one thing in their head and tries to ask questions about that. It's very much human nature to fixate on one thing and try to prove and disprove it. A machine's ability to question and remain objective is so much better—it can hold 1,000 options in its head."

While a human, having established they are looking for an animal, might try to find out if it is a reptile, amphibian, mammal, fish, insect, or bird, say, a more efficient question might be to find out if it lives in the sea or above the sea. That way, whatever the answer, you have gotten rid of about half the options. If you ask if it is an insect and the answer is no, you have only eliminated around a seventh, which is hugely less efficient.

Using a binary search, if people are drawing from 1,000 objects, in theory 10 questions should find the answer. With 5,000, it is 13. With twenty questions, you can guarantee correctly guessing anything from over one million objects.

In the real world this does not always work. Assuming you are not allowed to ask about its spelling ("Does it come between Aardvark and Mantis in the dictionary?") there is not always an obvious way to divide a group in two. A few simple tactics can help, though.

The first, which often trips up beginners, is to use comparators efficiently. If you ask, "Is it large?" unless it is either an atom (definitively small) or the universe (definitively big), almost any answer is correct. A stag beetle, for instance, is big for a beetle, but small for an animal. So instead ask if it is larger, say, than your head. Then there is no ambiguity, no wiggle room for a pedantic guessee, and you are able to create an arbitrary cutoff point—much like specifying Mantis in the above alphabetical example—through which you can divide the remaining objects in two and get closer to a binary search.

The second is to use words like *usually*. If you ask someone who is thinking of a sheep, "Would lions eat them?" the answer has to be Yes—lions, especially in zoos, have probably eaten sheep and, even if they haven't, given half a chance they certainly would. This will lead you down a cul-de-sac of asking about African herbivores. Ask instead, "Would lions usually eat them?" and the answer has to be No.

All of this assumes your guessee actually knows what they are talking about, mind you. This particular problem could be called "the dolphin constraint." Robin, whose computer learns from guesses rather than encyclopedias, is familiar with the issue. "I can ask our program, is a dolphin a fish? It's pretty sure it is a fish. Most of the people playing the game think that a dolphin is a fish." So even if you were able in theory to eliminate half of all options at each stage, you might find at the end that when you had carefully removed all the mammals, the dolphin had persisted: a Linnaean taunt, in parlor game form.

There are two ways to respond to this. One, fifteen questions on when you are trying to work out how a turbot could have starred in a hit 1960s drama series called *Flipper*, is to declare yourself the victor and the guessee an idiot. This is how many games, and friendships, end in the real world.

Robin's website has a more liberal approach. "The program is not an authority on knowledge, it is an authority on human knowledge. We call it folk taxonomy." It accepts that there is a 60 percent chance of a dolphin's being a fish in the mind of the guessee and never quite eliminates it. Which seems like a far more sensible approach. After all, what could be more satisfying than defeating your friend at twenty questions and then adding to the triumph by explaining to them that a dolphin is a mammal?

FACT

In the 1940s, twenty questions proved a popular format for a US radio show of the same name. Listeners had to submit the subject. The most frequent submission? Winston Churchill's cigar.

EPILOGUE

I learned the true meaning of a family Christmas when I was eleven or twelve. Gathered with my relatives in Ireland, we were playing a board game, I forget which. What I do remember is what happened when I lost, and my uncle Terry won.

Sitting in front of a blazing fire, glasses of port for the adults and comforting hot chocolates for the children, Terry looked at me—a nephew he had known since the day I was born, who had played with his own children and spent holidays running in his garden.

Then, his performance began.

"Ha ha, I won, you lost," he sang (yes, he actually sang). He leapt on the sofa and bounced up and down, with each bounce pointing at me and singing "you lost" for emphasis. Finally, he sat down, brushed himself off, regained his composure, and asked for another game. No one passed comment.

Terry was far from the only uncle who assisted in the creation of this book. There was Uncle Ken, who structured his year around his annual Christmas quiz. In his case the pleasure was not in winning—he was the quizmaster—but in devising questions so devious that we were all losers. Then there was Uncle Simon, who introduced me to Diplomacy, a game few people have ever won, but whose sine qua non is forming friendships and alliances with more trusting nieces and nephews, then brutally doublecrossing them after twelve hours of game play.

It is always invidious to single out people in an acknowledgments section, but equally it would be wrong not to mention the important role my cousins and my own sister had in forming the philosophy behind this book. It was my cousin Olivia who first introduced me to the delights of daylong games of Monopoly. It was her brother, Christopher, often in alliance with my sister, Clare, who showed me how such games ordinarily end: with someone tipping up the board.

But my final lesson in gamesmanship, one that I save for another book, came not from relatives but from my neighbors. For much of my childhood we lived next door to a gang of unruly children named the Salisburys. They taught me that, ultimately, the real victory comes not on the game board at all; that is merely a prelude. The real victory comes to whoever can afterward put his opponent in a headlock, wrestle him to the ground, and then sit on his face.

Uncle Terry: watch your back.

ACKNOWLEDGMENTS

It is always the case that a lot of people are involved in the creation of a book. That is especially true here. Each chapter required the time, effort, and experience of experts in the field. Not only were they kind enough to talk to me about their jobs and lives, they were also kind enough to do so in the cause of a book about games.

One of the privileges of my day job as a science writer is that I get to chat to people about their life's work and distill it into 600 words. I long ago realized that someone who is passionate about quantum mechanics, or snails, or stone skipping, is a hugely more interesting interviewee than an A-list actor or a celebrity chef. That theory was proved here.

My only regret is that two experts were left out: Daniel Sheppard on Rubik's cube and Michael Bowling on poker. With both the fault was mine. They were engaging and knowledgeable—it is difficult to think, in Sheppard's case, of how I could have found someone more qualified than a man who can solve a Rubik's cube blindfolded, in under a minute—but my own talents were lacking: I was unable to transfer their expertise into something comprehensible on the page.

Jamie Joseph at Ebury was all I could have hoped for in an editor: just enough praise to keep my fragile ego happy, and enough gentle suggestions for structural improvements hopefully to keep readers happy, too. He is responsible, along with designers at Clarkevanmeurs Design and illustrator Tiffany Beucher, for overseeing the beautiful final look of the book. Helena Caldon helped get rid of some of the sillier mistakes. Both she and Jamie should be thankful to my wife, who saved a lot of editing and probably a lot of ego massaging, too, by eliminating the more self-indulgent passages and clunkier phrases. Without Sarah Williams this book simply would not be here. She is the perfect agent: efficient, kind, wise—and she doesn't wear red trousers or drink Nespressos.

Finally, I would like to thank *The Times* of London. Early in my career one of the editors gave me a piece of advice. He said, "Whenever important people agree to interviews or invite you to parties, remember they are not talking to you at all. They are talking to *The Times*."

I owe everything to the paper. Far too many people at Times Towers have helped me get to the stage where I could pitch a book and have a publisher notice for me to name them all, but it would be wrong not to mention several: Ben Preston, Martin Fletcher, and Bronwen Maddox, who gave me my first job; Roland Watson and Emma Tucker, who helped me get my second; and Nicola Jeal, who for a while kept me occupied every Friday finding trends in male beauty products, Great White Sharks in Cornwall, and even self-awareness in Jedward.

It was James Harding who moved me to science, and John Witherow, Fay Schlesinger, and Jeremy Griffin who decided to promote someone whose milieu is research papers about seals having sex with penguins and why dogs defecate in a north-south direction. I am hugely grateful. It is the home news desk—Dan Parkinson, Mark Sellman, James Burleigh, Andrew Ellson, Devika Bhat, Claire Bishop, Dee Howey, and Robin Stacey—who have to live with the daily consequences of that decision. Lastly, massive thanks to Mike Smith and Michael Moran.

Any mistakes are mine and mine alone.

APPENDICES

APPENDIX 1: QUIZ ANSWERS

1. Aldrin   2. Saint Barthélemy   3. Arachne

APPENDIX 2: TWO-LETTER SCRABBLE WORDS

All words are in Merriam-Webster's Official SCRABBLE Players Dictionary (4th Edition). Lists compiled by the North American SCRABBLE Players Association (www.scrabbleplayers.org).

aa   (Hawaiian) a type of lava > AAS.

ab   An abdominal muscle > ABS.

ad   (Coll.) advertisement > ADS.

ae   (Scots) one. No -S.

ag   (Short for) agricultural; (noun) agriculture > AGS.

ah   Interjection expressing surprise, joy, etc. > AHS, AHING, AHED.

ai   (Tupi) the three-toed sloth > AIS.

al   An E. Indian shrub > ALS.

am   Present tense of be.

an   Indefinite article; (noun) something that might have happened but did not, as in ifs and ans > ANS.

ar   The letter r > ARS.

as   In whatever way; (noun) a Norse god > AESIR; a gravel ridge or KAME > ASAR; a Roman coin > ASSES.

at   Preposition denoting position in space or time; (noun) a monetary unit of Laos > ATS.

aw   Interjection expressing disappointment, sympathy, etc.

ax   (US) axe > AXES.

ay   (Noun) an affirmative vote > AYS. Also AYE.

ba   In Ancient Egyptian religion, the soul > BAS.

be   To exist.

bi   (Short for) bisexual > BIS.

bo   Fellow; pal, buddy > BOS.

by   Beside, near; (noun) same as BYE > BYS.

de   From (as used in names).

do   A musical note: DOS; (verb) to perform > DOES, DOING, DID, DONE.

ed   (Short for) education > EDS.

ef   The letter f.

eh   Interjection expressing enquiry; (verb) to say "eh" > EHS, EHING, EHED.

el   The letter l > ELS.

em   The letter M; a unit of measurement in printing > EMS.

en   The letter N; a unit in printing > ENS.

er   An interjection expressing hesitation.

es   The letter S > ESES. Also ESS.

et   (Obs.) pt. EAT.

ex   The letter X; someone no longer in a previous relationship > EXES; (verb) to cross out > EXING, EXED.

fa   A musical note, as in sol-fa > FAS.

fe   (Hebrew) a Hebrew letter > FES.

go   To pass from one place to another > GOES, GOING, WENT, GONE; (noun) a board game > GOS.

ha   An interjection expressing, e.g., surprise.

he   A male person > HES.

hi   An interjection calling attention.

hm   An interjection expressing hesitation. Also HMM.

ho   Interjection calling attention, expressing surprise, etc. Also HOH.

id   A fish of the carp family > IDS. Also IDE.

if   On condition that; (noun) a condition > IFS.

in   (Verb) to take in > INS, INNING, INNED.

is   (3rd.) BE, to exist.

it   The neuter of he, she, him or her.

jo   (Scots) a loved one > JOES.

ka   The spirit or soul of a dead person; (verb) to serve > KAS, KAING, KAED. Also KAE.

ki   (Japanese) the spirit of Japanese martial art > KIS. Also QI, CHI.

la   A musical note > LAS.

li   (Chinese) a Chinese unit of distance > LIS.

lo   An interjection meaning see, look.

ma   (Coll.) mother > MAS.

me   A musical note > MES.

mi   A musical note > MIS.

mm   An interjection expressing agreement.

mo   A moment > MOS.

mu   (Greek) a letter of the Greek alphabet > MUS.

my   Of or belonging to me.

na   (Scots) no, not at all.

ne   (Obs.) not.

no   Word of negation > NOS or NOES.

nu   (Greek) a letter in the Greek alphabet > NUS.

od   A hypothetical force; an old word for god, often used as a mild oath > ODS.

oe   (Scots) a grandchild > OES. Also OY, OYE.

of   Belonging to.

oh   An interjection; (verb) to say OH > OHS, OHING, OHED.

oi   An interjection used to express attention; (noun) the gray-faced petrel > OIS.

om   An intoned Hindu sacred symbol > OMS.

on   (Verb) to go on with, to put up with > ONS, ONNING, ONNED.

op   (Short for) operation > OPS.

or   (Noun) the heraldic tincture gold > ORS.

os   (Lat.) 1. a bone > OSSA. 2. a mouth-like opening > ORA.

ou   (Scots) an interjection expressing concession; (noun) a bloke > OUS.

ow   An interjection expressing pain.

ox   A bovine animal > OXEN; also, a clumsy person > OXES.

oy   (Scots) a grandchild > OYS. Also OE, OYE.

pa   (Maori) a hill fort > PAS. Also PAH.

pe   (Hebrew) a Hebrew letter > PES. Also PEH, FEH.

pi   (Greek) a letter in the Greek alphabet > PIS.

qi   (Chinese) the physical life force postulated by certain Chinese philosophers > QIS.

re   A musical note > RES.

sh   An interjection requesting silence. Also SHA, SHH.

si   An earlier form of TI, a musical note.

so   In such a way; (noun) a musical note > SOS.

ta   An interjection expressing thanks > TAS.

ti   A musical note; a small Pacific tree > TIS.

to   In the direction of, toward.

uh   An interjection expressing surprise.

um   An interjection expressing doubt or hesitation; (verb) to express hesitation > UMS, UMMING, UMMED.

un   (Dial.) one > UNS.

up   (Verb) to move up > UPS, UPPING, UPPED.

us   Pronoun.

ut   A musical note > UTS.

we   Pronoun.

wo   Woe > WOS.

xi   (Greek) a letter in the Greek alphabet > XIS.

xu   A Vietnamese monetary unit > XU. No -S.

ya   You. No -S.

ye   (Arch.) you.

yo   An interjection calling for effort or attention.

za   (Sl.) pizza > ZAS.